Environmental Toxicology

John H. Duffus

Department of Brewing and Biological Sciences, Heriot-Watt
University, Edinburgh

Edward Arnold

© J. H. Duffus 1980

First published 1980 by
Edward Arnold (Publishers) Ltd
41 Bedford Square, London WC1B 3DQ

British Library Cataloguing in Publication Data

Duffus, John J
    Environmental toxicology. — (Resource and environmental
      sciences series).
    1. Pollution — Toxicology
    I. Title II. Series
    614.7 RA565

ISBN 0-7131-2798-8

Printed in Great Britain by
Spottiswoode Ballantyne Ltd.
Colchester and London

Preface

This book reflects my belief that the analysis of environmental effects of toxicants requires an interdisciplinary approach. This approach is developed from a first chapter dealing with assessment of toxicity through chapters describing the metabolism of toxicants by animals, plants and micro-organisms, and four chapters discussing our knowledge of toxicants of current concern. Finally, the problems of utilizing this knowledge in the environmental context are considered.

Throughout this book, the content of mathematics, physics and chemistry has been kept to a minimum, in order to make the contents accessible to the widest possible readership. Although intended for students, it should also be useful to qualified scientists and other specialists who find themselves faced with problems caused by environmental toxicants. Such specialists may find the book more useful as a work of reference than as a text and, to facilitate such usage, a comprehensive index has been compiled, including as many as possible of the proprietary synonyms of the commercial toxicants which are mentioned. Full chemical names of these toxicants are given in the Appendix. Non-biologists may find a biological dictionary useful to explain unfamiliar biological terms. A bibliography is given at the end of the book to guide the reader to the relevant literature, and a brief general guide to relevant information sources is given in the Appendix.

In writing this book, I have drawn heavily on my experience in teaching environmental toxicology a the Heriot-Watt University, and I am indebted to the students who have helped me clarify my ideas. Apart from the students, I am especially indebted to my colleague, Cliff Johnston, for suggesting to me that environmental toxicology was a subject suited to my general scientific interests.

This book was written at the University of Massachusetts at Amherst during a visit made possible by permission of Professor D. J. Manners and Professor F. W. Southwick. I greatly appreciated the help given to me by Paul Jennings, Herb Marsh and many other faculty and staff at Amherst. Last, but by no means least, I am grateful to my wife, Carol, for reading through my manuscript and eliminaing many errors and to Mrs Faye Craig for producing the typescript.

Edinburgh   J. H. Duffus
1980

Contents

1 Assessment of Toxicity

Environmental toxicology is the study of the effects of toxic substances occurring in both natural and manmade environments. The main task of the environmental toxicologist is to assess objectively the risks resulting from the presence of such substances. On the basis of this assessment, the environmental toxicologist may be asked to advise on measures to prevent the substances reaching harmful levels, or to minimize damage where harmful levels have already been attained. Whatever advice is given, it may have economic implications leading to considerable expenditure, either from private or public funds. It is therefore essential that the assessment of risk be made in full knowledge of the potentialities and the limitations of the procedures likely to be used. In this chapter the methods most commonly favoured will be surveyed and the significance of the results obtained discussed. Laboratory studies must be correlated to field observations and the problems involved in this will be considered where appropriate.

1.1 Assessment of Short-term Lethality and Acute Toxicity

The most commonly used assessment of toxicity is the measurement of short term lethality. For a given substance, this involves determining the median concentration which is lethal to a fixed proportion, usually 50%, of a test population of organisms after continuous exposure for a fixed time, usually 48 or 96hrs. This is interpolated from the sigmoidal dose response curve (Fig. 1.1) which is obtained by plotting percentage mortality at each dose against dose. The use of probits to linearize the relationship may be advisable, and is fully described in Finney, *Probit Analysis* (1971).

The lethal concentration (LC) for 50% of a population after continuous exposure for 48h is referred to as the 48h LC_{50} or LD_{50} (LD = lethal dose). This abbreviation may be altered appropriately to correspond to different exposure times and proportions of the population. Usually the concentration referred to is that in aqueous solution but it may also be a concentration in air. If the test compound is insoluble or sparingly soluble in water, it must be uniformly dispersed for a repeatable LC_{50} to be obtained. If an emulsifier or other solubilizing agent is used for this purpose, it should be chosen carefully to ensure a minimal contribution to the toxicity of the system.

Difficulties arise where organisms are exposed to harmful substances in association with particulate matter, or in their prey or other food. What then is the effective concentration to which they are exposed? The nature of the particulate matter will determine whether the substances are ingested or not. It will also determine the potential for dissociation of toxic substances

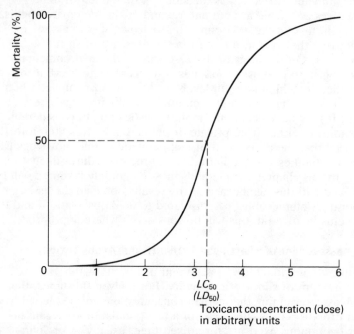

Fig. 1.1 Graph showing typical relationship between mortality and toxicant concentration, or dose, following continuous exposure of organisms for a short fixed time, usually 48 h or 96 h.

following ingestion. Normally, only dissociated material can exert toxic effects. In a food organism, toxic substances may be converted to derivatives of greater or less toxicity and may be localized in specific parts of the organism, some of which may be selectively eaten or discarded by its predators. Therefore, in these cases it is almost impossible to know the effective concentration to which the predator is exposed. The best one can do is to copy the natural situation as closely as possible in the laboratory and express the LC_{50} as a notional concentration, taken as the total available toxic material divided by the weight of particulate matter or by the weight of the food organism. If one knows that a certain derivative of the toxic substance is much more lethal than any of the others, and suitable methods of analysis are available, the derivative concentration in the food organism should be determined since this will be the most significant factor. In medical usage the LC_{50} frequently refers to an injected concentration, and this must be borne in mind when attempting to extrapolate from such studies.

Lethal concentrations have been expressed in a variety of units, most frequently in milligrams per litre or per kilogram body weight, but it is desirable that a more uniform approach should be adopted. Where possible, molarity (moles per litre) or molality (moles per kilogram) should be used as this will give a uniform chemical basis for comparing toxicity. The value of the LC_{50} as a measure of toxicity has been queried, and it is important to realize that it may be readily misinterpreted. Firstly, it is not constant. Considerable variation is observed, even in a given laboratory with the same population under the same conditions and the same experimenter. Secondly, there is no simple correlation between organisms permitting direct extrapolation, even between organisms of the same Class such as rats and man. Thirdly, it gives no indication of the concentrations at which sublethal harm may occur. It is quite possible that a toxicant may be lethal only at very high doses but produce significant ill effects at relatively low concentrations. The most information the LC_{50} gives is an idea of the order of magnitude of the lethal dose under specified conditions. However, it is relatively quick and cheap to determine and provides a basis, however rough, for initial assessment of the likely hazard from a toxicant and of the effects of various parameters on its toxicity.

Apart from the LC_{50}, a number of other figures may be derived from studies of short term lethality. For example, the minimum lethal dose (MLD) is the dose which will kill at least one organism in the test population over the test period. This may be described as the LD_1. One may also determine, as Pham-Huu-Chanh has suggested, the maximum does never fatal in 24h ($MDNF$) and the minimum dose always fatal in 24h ($MDAF$) which is the smallest dose causing 100% mortality in that time. Another possibility, suggested by Luckey and Venugopal (see Chapter 6), is that the concept of potential toxicity (pT) be introduced on an analogy with pH, i.e. $pT = -\log[T]$ where $[T]$ is the molal concentration of the toxicant and the logarithm is to the base 10. A toxicant with a 24h LD_{50} of 0.001 mol kg^{-1} would have a 24h $[T_{50}]$ of 10^{-3} and a 24h pT_{50} of 3. The calculated pT corresponds directly to the effect of toxicant on the test population. A small pT_{50} should indicate a relatively harmless substance while a large pT_{50} will indicate a highly toxic substance, at least in the context of lethality in the test system.

Measurement of short term lethality is one aspect of assessment of acute toxicity. Acute toxicity has been defined most recently as 'the total adverse effects produced by a toxicant when administered as a single dose' (Hunter and Smeets, 1977) and this will be the definition applied here. However, alternative definitions are used, e.g. 'the total adverse effects produced by a toxicant when given in a single dose or multiple doses over periods of 24 hours or less'. Therefore, care must be taken to understand the terminology used by a particular author before any appraisal of results can be undertaken. It will be seen that both definitions quoted refer to total adverse effects which emphasizes the point that there may be many harmful effects before lethality supervenes, and all effects of the toxicant must be monitored throughout testing. Further, thorough *post mortem* examinations must be carried out on any organisms that die. In this context another parameter may be defined, i.e. the EC_{50} or ED_{50}, which is the concentration or dose which produces a specific ill effect in 50% of the organisms tested. The EC_{50} is subject to the same provisos given for the LC_{50}. It will be seen that the ideal result of acute toxicity testing will be a table containing the LC_{50} and a range of EC_{50} values covering all possible effects.

An alternative approach to the measurement of acute toxicity is to measure the ET_{50} which is the median exposure time to a given concentration of the toxic substance required to produce 50%

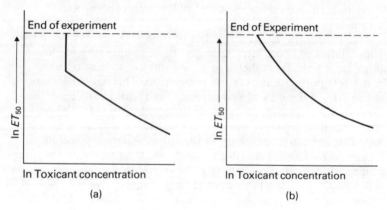

Fig. 1.2 Possible relationships between ET_{50} and toxicant concentration: a, graph indicating the existence of a safe threshold concentration; b, graph suggesting that the toxicant will be lethal at all concentrations, given sufficient exposure time.

mortality in a test population. This is measured over a range of concentrations and requires continuous monitoring. It therefore involves much more laboratory effort but it can yield more useful information than the 48h LC_{50}. For example, a plot of the ln ET_{50} against ln concentration of toxic substance may indicate a threshold concentration below which the substance is not lethal ((Fig. 1.2a). However, it may indicate that the substance is lethal at all concentrations, low concentrations simply requiring long exposures to take effect (Fig. 1.2 b). While such plots may be strongly suggestive, they cannot be regarded as conclusive evidence. Thus, where there appears to be a threshold in these experiments, which are of fairly short duration, long exposures may lead to significant mortality due to accumulation of the toxic substance. Similarly, an abrupt threshold may appear in a plot which otherwise displays a continuous trend.

1.2   Assessment of Long-term Lethality and Chronic Toxicity

Tests of short-term lethality and acute toxicity have long been used to assess risk, but in many environmental situations the observed problems are caused by the response of organisms to concentrations of toxic substances which are harmful only after long periods of continuous exposure. One consequence of such exposure may be death, but the identification of such death as long-term lethality from the toxicant is difficult to establish. The death may be due to natural causes, enhanced or accelerated by the weakened state of the test organism but not directly due to the toxicant. Further, the longer the time over which lethality is determined, the greater the variability of the results. All the criticisms of the short-term LC_{50} also apply, so long-term lethality is not normally considered a useful parameter for measurement.

By analogy with the definition of acute toxicity, chronic toxicity may be defined as 'total adverse effects produced by a toxicant when administered continuously over a long period of time'. Assessing chronic toxicity is not easy, primarily because of the practical difficulties in maintaining organisms under constant conditions in the laboratory for long durations. The fundamental problems are the same as for acute toxicity tests but the time factor greatly exacerbates them. While almost all laboratory toxicity assessment has been based on continuous exposure of organisms to a fixed concentration of toxicant, it is highly unlikely that this situation ever occurs in nature. Therefore, there is a good case to be made for setting up experimental systems where toxicant concentration fluctuates in a controlled manner, particularly where chronic and sublethal effects are being assessed. However, the difficulties in setting up such systems will be considerable and it will generally be more convenient to obtain similar information by careful and thorough monitoring of toxicity responses in the field as they arise.

1.3   Practical Problems in Assessing Toxicity

1.3.1   General considerations

The first practical point to be considered in carrying out toxicity tests is the nature of the test population. Ideally, the organisms used should be genetically identical and pathogen free. They should be kept under sterile conditions in a constant environment and lit with artificial light, as nearly as possible equivalent to sunlight, for a twelve hour day. These are the requirements of the toxicologist whose main concern is a thoroughly repeatable assay of toxicity. Whether these should be prime requirements of an environmental toxicologist is open to question. By definition, the environmental toxicologist is concerned with the risk to natural populations which are genetically heterogeneous, subject to the effects of pathogens and living in a variable environment. It may be that one important effect of a toxic substance will be to eliminate selectively sensitive individuals from a population. Another effect may be to eliminate individuals affected by a specific pathogen. Other effects may only become apparent under extreme environmental conditions. There are many possibilities, none of which are covered by the analytical toxicologist's ideal population. This is not to say that the carefully selected and maintained populations of organisms, such as inbred rats and mice, which have been developed by medically-orientated analytical toxicologists over the years, do not have their place in environmental toxicology. They are useful for the bio-assay of minute amounts of toxicants, especially where the chemistry of these substances is unknown or does not permit sufficiently sensitive chemical analysis. They may also give pointers to effects on other organisms. Direct extrapolation cannot be justified because of the known magnitude of species and generic differences.

1.3.2   Selection of organisms and their maintenance

In selecting organisms for toxicity assessment, the environmental toxicologist must bear in mind the relevance of the organisms to the environment of interest. The effects of air pollution on plants may be monitored using lichens, which show high sensitivity to many toxicants (see Section 3.8). Other plants may be used where appropriate, a limiting factor frequently being ease of maintenance in the laboratory. This limiting factor is particularly important in selecting animals for testing. Another important limiting factor is ignorance of the normal characteristics and degree of variability of many organisms. There are very few organisms for which there is sufficient fundamental knowledge relating to their life under controlled laboratory conditions, far less under the conditions of their normal habitat.

Assuming that a suitable organism can be found, i.e. of demonstrable importance in the environment of interest, with a good background of previous study and easily maintained in the laboratory, a breeding population under suitable defined conditions can be established. A breeding population is essential because susceptibility to toxic substances alters with the developmental stage, typically being greatest in embryonic and larval stages, and in senescence. Suitable defined conditions mean that, until the direct toxicity on the organisms has been assessed, the population should be kept pathogen free as far as possible. Adequate nutrition of known composition must be supplied. In most cases this will be maximal nutrition, with food intake being controlled only by the organism. Anything less will lead to competition between individuals for the available food resulting in a high degree of variation in food intake throughout the population. Maximal food intake is not the same as optimal food intake and it is certainly not the most usual occurrence in any natural environment. However, it does provide a defined base from which to work. Other environmental factors to be considered are light, temperature, humidity, daylength, nature of living space (which may have behavioural effects), subjection to handling, noise or other disturbances, and, for aquatic organisms, pH, salinity, degree of oxygenation and other parameters of water quality.

Initially, these factors will be established primarily to suit the experimenter's convenience, but the long-term aim must always be to approximate as closely as possible to the particular environment under consideration. This ultimately requires complex, and expensive, control systems and, possibly, an even more complex computer program to analyse the results obtained. At this point, the laboratory experimenter would be wise to consider the possibilities of a properly planned field study. If a suitable area has already been contaminated with the toxic substance of interest, a study in depth should be carried out, monitoring all possible parameters, and the results should be correlated with those from the laboratory. If no contaminated area is available, setting up a field experiment must be done with great care. The experimental area must be totally isolated from the surrounding environment so that the toxic substance does not become a general hazard and so that the experimental area can be readily decontaminated at the end of the experiment. The experimenter should also be familiar with any legal requirements that may apply. In setting up a field experiment of this kind much useful information can be obtained if part of a laboratory population of organisms is included. The remainder of the population can then be subjected to similar conditions, at least with regard to the concentration of the toxic substance, in the laboratory. This should improve correlation of laboratory results and field observations.

1.3.3   Exposure to the toxicant

Having selected the organism and provided it with suitable living conditions, it can then be exposed to a range of concentrations of the toxic substance. Having made sure that the substance is pure to start with, it is crucial to know the exact concentration. Some compounds are inherently unstable and, if this is the case, the breakdown products should be considered. If they are more toxic than the original substance, it would be better to study them. If much less toxic, continuous addition of the unstable compound to the test system may be necessary to maintain a constant concentration and level of toxicity. However, if the unstable substance is normally released into the environment in pulses with an appreciable time lag in between, a measurement of toxicity at constant concentration might not have much relevance. Assuming that it has, even with chemically stable substances, maintenance of a constant concentration poses problems. Uptake of the test substance by the test organisms and subsequent metabolism will lower its concentration. Other organisms in the system, especially micro-organisms, may have the same effect. Chemical trans- formations such as photodecomposition may occur. Adsorption of the test substance on to containers and tubing may be appreciable. Adsorption on to particulate matter will reduce the concentration in solution but may facilitate ingestion of adsorbed compounds by filter feeders such as oysters or mussels. Volatile compounds may be lost by evaporation and exert different toxic effects in the gaseous phase. In aqueous solutions surface active compounds will accumulate at the surface. Where possible, these problems may be reduced by continuous replacement of the relevant gaseous or aqueous phase. Substances which do not mix well with the gaseous or aqueous media may cause difficulties since the concentration in contact with the organism may not be the same as the overall concentration. The precautions to be taken when using adjuvants to aid mixing have already been referred to, but it should be remembered that the alternative of constant agitation may also have adverse effects. The one essential measure to be taken when toxicity testing is being carried out is to monitor the concentration of the toxic substance routinely throughout the tests. Then, even if the concentration is not constant, at least effects can be related to known concentration changes.

Apart from the concentration of the toxic substance, other factors to be considered are those which may increase toxicity, i.e. synergistic factors and those which may decrease it, i.e. antagonistic factors. These factors may be other chemicals which enhance or reduce the observed toxicity (see Chapter 2) or they may be environmental or biological factors. Generally, conditions unfavourable to survival of the test organism will be synergistic while favourable conditions will be antagonistic, but more subtle interactions are possible. Biological factors should already have been taken into account in selecting the test population and the importance of developmental stage and physiological condition should be remembered. Connected with this is a tendency for the susceptibility of organisms to decrease with increasing size.

1.3.4 Assessment of chronic toxicity

Chronic toxicity may be assessed in much the same way as acute toxicity, given that the practical difficulties can be overcome. A major difference is that chronic effects are often less directly related to environmental concentrations and are better related to levels accumulated in the organism or even in a specific tissue. This gives rise to the concept of the 'critical organ' or 'target organ'. The critical organ is best defined as that organ which first shows harmful functional changes under specified conditions of exposure for a given population. It should be noted that this is not necessarily the organ of greatest accumulation, e.g. lead accumulates mainly in bone where it causes little evident harm. Also the critical organ may vary. The lung is the critical organ for short-term exposure to high levels of atmospheric cadmium but the kidney is the critical organ for long-term exposure to low levels. Having identified the critical organ, the critical concentration of the toxicant, i.e. the mean concentration in the critical organ at which adverse effects become detectable, may be determined. Further, subcritical effects and corresponding subcritical toxicant concentrations may be defined as being those which produce detectable changes which do not impair organ function. This may be useful in providing an early warning system for prevention of harmful effects.

Where accumulation of a toxicant occurs in a noncritical organ, an organism may be put at risk by subsequent exposure to stress. For example, chlorinated hydrocarbons accumulate in adipose tissue which can be rapidly broken down in animals subjected to stressful conditions. The subsequent release of the chlorinated hydrocarbons into the bloodstream may have serious effects, approximating to those observed in acute toxicity tests, though resulting from chronic exposure. Thus, the distinction between chronic and acute toxicity may sometimes be blurred.

Finally, it should be remembered that even a toxicant which itself is not accumulated may have cumulative effects. This would be a very difficult relationship to establish but it should not be ignored.

1.3.5   *Assessment of sublethal effects*

Lethality is a very crude measure of toxic response. Therefore, there has been a search for sublethal effects that might be monitored. These effects would occur at much lower concentrations of the toxic substance than those causing death and so might provide a very sensitive means of detecting its presence. Further, since it is likely that most toxic substances in the environment will be present at sublethal levels in most circumstances, it becomes imperative to know when these levels should be regarded as dangerous. In this context, one must remember that not all changes resulting from exposure to a toxic substance are necessarily harmful (Fig. 1.3). Some will simply be part of normal homeostasis as an organism adapts to its changed environment. Eventually a concentration will be reached where the homeostatic mechanisms will be so stressed that the organism will cope only with difficulty. Clearly the concentration must be kept well below this level if the organism is to be protected. The level causing undue stress will vary from organism to organism and so, in establishing this level for a given ecosystem, it will be essential to find the most sensitive organism present.

Parameters of sublethal toxicity that have been suggested for measurement range from the biochemical to the behavioural. Any toxic compound will probably affect a number of these parameters and for each there will be a different response curve. Each parameter must be studied individually and no extrapolation from one to another can be permitted unless a relationship has been rigorously demonstrated. Ideally one would like to be able to discover parameters which are altered only by specific toxic

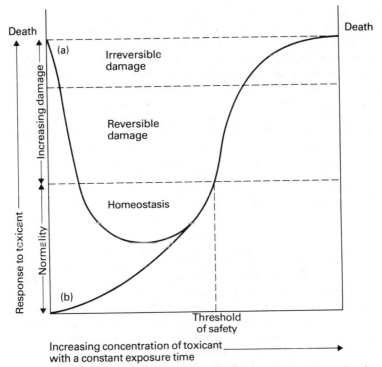

Fig. 1.3 The relationship between observed effects and toxicant concentration, i.e. dose-response curve: a, the observed effects of a substance essential to life at low concentrations; b, the effects of a nonessential substance.

substances but these are rare. One such, at the biochemical level, is the inhibition of acetylcholine esterase by organophosphorus insecticides. This can be used to detect the effects of these insecticides, but it must be remembered that the point at which the inhibition goes beyond the homeostatic capacity of an affected insect is not known. In fact, it is still very difficult to deduce changes at the organism or population level from knowledge of effects at the biochemical level. Amongst the most convenient biochemical changes to measure in animals are those affecting the blood. Blood samples can be obtained from many animals quite readily over a period of time without any marked adverse effects. However, taking blood samples inevitably involves handling, with consequent stress to the animal, and allowance may have to be made for this. Techniques for blood analysis have been highly developed for use in medical examination of human beings and there is usually little difficulty in adapting them for use with blood from other organisms. The changes most commonly monitored are those in enzyme activities, serum proteins and electrolyte concentrations. Note should also be taken of any gross changes in the blood cells visible on routine microscopy. This leads on to the consideration of gross pathological changes in essential organs but these have the disadvantage that they may require the sacrifice of the organism or, at least, some kind of surgical intervention. Monitoring of some kinds of physiological change can be done in a nondestructive fashion, e.g. measurement of oxygen consumption but, as with biochemical changes, it may be difficult to define the difference between homeostatic and pathological changes.

It is unlikely that any single parameter can ever be entirely satisfactory for monitoring sublethal effects. Therefore, it would seem desirable to monitor as many as possible in any given situation and to try to assess the overall picture. This was the approach adopted by Eisler in compiling what he called 'stress profiles' for the mollusc *Venus mercenaria* exposed to methoxychlor and malathion. Eisler demonstrated a number of metabolic changes but it was not clear whether these were harmful as there were no obvious deficiencies in feeding or reproduction. The alternative approach is to monitor something which represents an integration of many biochemical and physiological processes. One possibility is behaviour. The problem with this is finding a behavioural response which is reasonably easy to measure and which has demonstrable significance for the survival of the organisms of interest. Another possibility is the laboratory production index as defined by Mount and Stephan. This is the change in biomass of a population measured over at least one generation. It incorporates effects on growth and reproduction, both of which are important for population success. The main drawback in this measurement is the length of time required for the experiment but, if it is possible, it should be well worth the effort.

1.3.6   *Assessment of effects on* Homo sapiens

So far the effects of toxic substances in the environment on organisms other than man have been considered but, inevitably, effects on man must be among the main concerns. Human beings cannot be used in tests of lethality. Information on lethality must be obtained from study of those human beings who have been exposed accidentally to lethal concentrations of toxicants. Similarly, sublethal effects can rarely be studied in properly planned experiments though, occasionally, such experiments may be possible if volunteers are available. Hence, risks to human beings are usually assessed using laboratory populations of other mammals, e.g. rats, mice, rabbits and dogs. Extrapolation from effects on these species to predict effects on *Homo sapiens* is no better justified than any other extrapolation between species. As a result, mistakes have been made in permitting human exposure to toxic substances following evidence of low toxicity in test animals. Similar mistakes may have been made in banning substances toxic to one or other of the common test species. However, there can be little argument that the consequence of mistakes of the latter kind are less to be deplored than the former. Mistakes of both kinds may be avoided to some extent by epidemiological studies. Analysis of the geographical or social distribution in the human population of defined toxic responses, or of the relation of these responses to medical treatment or other activity, may show a significant correlation with a specific toxicant. Conversely, no significant correlation with a suspected causal factor may be obtained.

Those involved in the production or application of toxicants are inevitably subjected to higher doses than are the general public. While data on their health may be on record with their work's health team, it is not always available for epidemiological study as this involves ethical problems and may require negotiations with management, medical officers and unions, because of potential legal consequences. Even if such information is available, it is rare for health monitoring to continue after employment ceases. This is most unfortunate because some ill effects, e.g. cancer, may not develop until several years after exposure. However, the idea of routine compulsory health monitoring of individuals and the retention of the results in a central record store is repugnant to many who feel that such information might be used against them. It might, of course, be of value to them in claiming compensation in cases where illness due to industrial toxicants could be demonstrated. It would certainly be of value to the general public for whom monitoring of such groups would provide an early warning system for dangers from industrial sources.

It should not be forgotten that the aim of much toxicity assessment is to set safe limits for levels to which human beings may be exposed (see Section 9.5). To do this, those members of the population who are most at risk and thus constitute what may be called the 'critical group' must be considered. Pregnant women will frequently come into this category because of the susceptibility of the developing foetus to toxic effects. However, there may be less obvious groups to consider, e.g. those who have a very high intake of foods where the toxicant occurs at unusually high levels. An example of this occurred in setting the safe limit for discharge of radioactive ruthenium (^{106}Ru) from the nuclear plant at

Windscale in England. This was established at a low enough level to protect Welsh eaters of laver bread which is made from a seaweed in which considerable ruthenium accumulation can occur (see Section 9.5.3).

1.3.7   *Assessment of teratogenic, carcinogenic and mutagenic effects*

As well as the effects already mentioned, teratogenic, carcinogenic and mutagenic effects must be given special attention. Teratogenic effects are those which cause malformation of the embryo. Carcinogenic effects are those which cause cancer, which may be defined as a disordered growth of cells which can invade and destroy tissues. The characteristic symptom is the formation of cell masses called tumours. Tumours which cause cancer are said to be malignant and those which do not are said to be benign. Tumours may occur in both plants and animals but cancer is found only in animals, possibly restricted to vertebrates. Mutagenic effects are those which cause chromosome alterations and thus alter genetic characteristics of cells. Toxicants which are mutagenic frequently exert teratogenic and carcinogenic effects in consequence.

To assess teratogenicity in the laboratory, a sensitive species must be found. Measurement of the LC_{50} may give some indication of sensitivity but there can be no guarantee that lethal toxicity is based on the same processes as teratogenicity. Having selected a species, it must be decided whether to use a strain which has a low spontaneous rate of malformations or one with a high spontaneous rate. With the former, very few abnormalities need be detected to exceed the background level and, therefore, only a small number of organisms will be required for testing. On the other hand, such organisms may be very insensitive to teratogenic effects. Hence, toxicologists have tended to prefer organisms with a moderately high spontaneous gross abnormality rate, usually comparable to the 1.5% rate that applies to *Homo sapiens*. This necessitates a fairly large number of test organisms to give statistically significant results.

Having selected the organisms, the route of administration of the substance under test must be decided. In many cases where animals are concerned oral administration will be chosen. If the toxicant is mixed with a carrier foodstuff the food composition must be accurately defined since it may affect absorption or subsequent metabolism of the toxicant. However, the toxicant may be in the water surrounding aquatic organisms or in air if an atmospheric toxicant is being studied. In every case, the route of administration should be selected to approximate to the normal environmental exposure. As well as the route of administration, the timing of administration may be critical. It is conceivable that sudden pulses of a toxicant at critical stages in embryo development may be more harmful than continuous exposure to the same level throughout development.

In studying mammals, foetuses which may be malformed should be obtained by Caesarian section one or two days before parturition as some mammals, especially rodents, eat foetuses which are born dead or malformed. However, normal parturition must be permitted in some of the exposed population since abnormalities may develop subsequent to birth. In mammals, the number of live and dead foetuses and/or resorptions should be recorded. All foetuses should be examined for cell, tissue, and skeletal abnormalities. In studying eyes, serial sections may be required to establish abnormalities.

Teratogenicity testing in organisms other than mammals is almost completely nonexistent. There is a clear requirement both for laboratory tests and for field surveys to establish whether early warning systems can be developed by studying sensitive species. An early warning system of this kind for detecting water-borne mutagens by monitoring chromosome mutations in ferns has already been described by Klekowski and Berger (1976). Another outstanding requirement is for a rapid, simple test of teratogenicity. Any test using higher organisms is extremely time-consuming and expensive and is unacceptable to the opponents of vivisection. Processes analogous to higher organism differentiation are known in microorganisms, e.g. spore formation, inter-conversion of unicell and hyphal forms and even changes through the cell cycle, but, to the author's knowledge, no-one has surveyed the effects of known mammalian teratogens on these processes. Bearing in mind the biochemical similarities among all living organisms, it seems possible that a microbiological test of teratogenicity could be derived from one or other of the processes mentioned. Those teratogens which are also mutagens stand a good chance of being detected by the Ames test (described on p. 16) or by related techniques.

As with teratogenicity testing, the first problem in assessing carcinogenicity is in selecting a suitable species. Similar criteria apply in respect of the LC_{50} as a possible measure of sensitivity and with regard to the spontaneous occurrence of tumours. the LC_{50} may not be related to carcinogenicity and low spontaneous tumour occurrence, while helpful in statistical analysis, may indicate low sensitivity to carcinogens. Both sexes of the selected species must be tested, as must all developmental stages through to senescence and death. The possibility of delayed effects is particularly important in relation to cancer and must be checked. The effects of a single dose of toxicant may become apparent only after a very long time. For example, it has been found that exposing a rat in mid-pregnancy to concentrations of N-methyl-N-nitrosourea which cause it no detectable harm can result in the offspring developing brain cancers on reaching maturity. This has been referred to as transplacental carcinogenesis.

Routes of administration of potential carcinogens must be chosen with regard to their environmental relevance, and the effects of superficial exposure in regard to skin tumours must not be neglected. Interactions of toxicants with other substances likely to be present simultaneously must be allowed for in design of the test system. Possible variations in response to continuous exposure or pulsed exposure should be assessed as far as possible. Allowance must be made for those substances called 'initiators', which are not necessarily carcinogenic themselves but which, when applied prior to a carcinogen, increase the occurrence of tumours or accelerate their appearance. There are also substances called 'promoters' which, when applied after a carcinogen, increase the incidence or accelerate the appearance of tumours. Assessing carcinogenicity with higher organisms is even more time consuming and expensive than assessing teratogenicity since the whole life span of the test organisms must be investigated.

A great step forward in recent years has been the development of microbial tests for mutagenicity which can be completed in a matter of days. Mutagenicity is the ability to alter the structure of deoxyribonucleic acid (DNA) in chromosomes causing a potential genetic aberration or mutation. Some of the many chemicals which have this ability are shown in Table 1.1. Cancer is essentially due to a metabolic disturbance in which growth and cell division of certain tissue cells ceases to be regulated in a normal fashion. Thus these cells proliferate to the detriment of the rest of the organism. It is now clear that in many, if not all, cases of cancer the metabolic aberration is the result of a mutation. Therefore, any mutagenic compound is a potential carcinogen. The most widely used microbial test for mutagenicity is that devised by Ames. The Ames system uses the bacterium *Salmonella typhimurium* as the test target. Strains of this bacterium have been selected which are unable to grow without histidine because they have a mutation in the gene group responsible for the manufacture of this amino acid. If the bacteria are exposed to mutagens, there is a high probability that the original mutation will be corrected in at least some cells. This process is called reversion. Once a bacterium has reverted, it can grow on histidine-free nutrient agar and form a colony which

Table 1.1 Common chemical classes of mutagens (adapted from *6th Annual Report of the Council on Environmental Quality (CEQ)*, CEQ, Washington DC, 1975).

Class	Examples	Typical sources
Aromatic amines	2-Naphthylamine	Manufacture of rubber products
	Benzidine	Dyemaking
Nitrosamines	Dimethylnitrosamine	Rubber additives
		Tobacco smoke
		Nitrosamines and nitrates in food
Chlorinated hydrocarbons	Vinyl chloride	Polyvinylchloride (PVC) manufacture
	Aldrin, dieldrin, DDT	Pesticides
	Bis-chloromethylether	Impurity in synthetic resins
Polynuclear aromatics	Anthracene	Tobacco smoke
	3,4-Benzpyrene	Combustion of fossil fuels
Radionuclides	Strontium-90	Medical, scientific and military use
	Plutonium-239	Nuclear power stations
Metal dusts	Arsenic	Pesticides and pharmaceuticals
	Beryllium	Lightweight alloys, rocket fuel
	Chromates	Paint pigments
Steroid hormones	Diethyl stilboestrol	Animal feed additive 'Morning after' pill

will be readily visible. The number of revertant colonies will be directly proportional to the potency of the mutagen.

In order to make this system more directly relevant to mammalian, and hence human carcinogenesis, a microsomal preparation (S-9) from rat or human liver may be added to the agar on which the bacteria are grown. Sometimes the test substances are pre-incubated with the extract. This extract carries out many of the metabolic conversions which may occur to chemicals. Some of these conversions may reduce their toxicity, others increase it (see Section 2.4). It has been calculated that the bacterial strains used in the Ames test can detect mutagenic activity of between 61% and 90% of known carcinogens. It has also been shown to give appropriate negative responses to chemicals that are believed to be noncarcinogenic. Thus, the Ames test has become widely accepted as a useful method for screening putative mutagens and carcinogens. It must be emphasized, however, that no single test can be considered as adequate for screening mutagenicity as false negatives may occur. Further, positive results in bacterial tests do not prove mutagenicity in other systems, nor does mutagenicity necessarily equate with carcinogenicity. Nevertheless, a substance shown to be strongly mutagenic in bacterial tests must clearly be regarded as dangerous to other organisms until proved otherwise.

Various criticisms have been made of the Ames test, primarily relating to the highly abnormal nature of the strains used and to the poor quantitation of the dose response relationship obtained. Hence, other short term tests have been developed. These range from bacterial tests in liquid culture (fluctuation tests) using the Ames strains, *Escherichia coli*, or *Bacillus subtilis*, through similar tests using yeasts, to tests using mammalian cells in culture. With mammalian cells in culture, mutagenicity testing is slower but the results should be more relevant to considerations of human health. Further, with some cell lines it is possible to monitor transformation, a process similar to carcinogenesis in which cells which normally grow only as a monolayer on a flat surface become able to form suspended colonies in soft agar.

Short-term tests can never eliminate the requirement for exhaustive testing with whole organisms and routine monitoring of organisms at risk. For whole organism testing, the currently favoured species are *Drosophila melanogaster* (the fruit fly), the rat and the mouse. The fruit fly has the advantage of thorough genetic characterization but the disadvantage that dosage of toxicants cannot be precisely controlled unless they are in the gas phase. Hypodermic injection is virtually impossible without severe injury and food intake is variable and not easily monitored. The rat may be used to detect chromosomal damage by examination of suitably stained bone marrow cells, or to demonstrate gross anatomical and physiological mutations, but the mouse is frequently the animal of choice for the latter. In either case, the tests are very time consuming and involve large numbers of animals and complex statistical analyses.

In addition to the above tests for mutagenicity, monitoring of accidentally exposed populations can often give valuable information. In eukaryotic organisms this generally means assessing chromosome damage. For human beings this is most conveniently done by taking blood samples and examining peripheral lymphocytes. However, visible chromosome damage requires a fairly massive effect of the mutagen, which at low doses may cause sister chromatid exchanges and gene mutations which are not detected by conventional techniques. Thus, negative results do not prove the absence of mutagens.

Another approach is possible if the putative mutagen can be obtained in a radioactively labelled form. Then the affinity of the chemical for DNA *in vivo* and *in vitro* can be quickly monitored.

Since all mutations represent alterations in DNA structure, any chemical which binds to DNA is likely to produce mutations. Again, however, negative results do not eliminate the possibility of the substance being a mutagen as indirect effects on DNA may occur.

1.3.8 The use of micro-organisms in assessing toxicity

The use of micro-organisms in toxicity testing has much to commend it. It is quick and inexpensive. Large numbers of test organisms are readily available under precisely controlled conditions so that results obtained have statistical validity and are readily repeatable. Additional advantages are to be gained from the detailed genetic and biochemical knowledge available about such organisms as *Escherichia coli* or, if a eukaryote is required, *Saccharomyces cerevisiae* or *Schizosaccharomyces pombe*. Problems arise, however, in extrapolating from the information obtained to deduce effects on more complex organisms. This will only become possible when toxic effects can be defined in relation to fundamental biochemical processes which are common to most organisms. To do this will require much further study, but the rewards in terms of rapid assessment of environmental hazards should make such study well worthwhile. Even without such information, the use of microorganisms in quantitative assays of known toxic substances, and in ranking of relative toxicities, ought to be perfectly feasible with considerable saving in expenditure on animal maintenance and reduction in animal suffering.

2 Metabolism of Toxic Substances by Animals

This chapter will be almost entirely devoted to a survey of our knowledge of the metabolism of toxic substances in mammals, since most relevant studies have developed from concern about the fate of drugs used in medical treatment of human beings. Fortunately, available evidence suggests that valid extrapolations may be made from the mammalian processes to analogous processes in other animals.

2.1 Uptake

The most probable route of uptake for many toxicants is from water or food through the tissues of the gastro-intestinal tract. Uptake may occur by diffusion, catalysed diffusion, or active transport.

2.1.1 Diffusion

Diffusion is the term used to describe movement down an electrochemical gradient, i.e. from a region of high concentration, either of molecules or electrical charge, to a region of low concentration or electrical charge. In relation to living cells this is often called passive transport and involves passage through the lipid bilayers of cell membranes. Hence, diffusional uptake of lipid soluble compounds is favoured. Lipid solubility is associated with absence of ionization. Accordingly, the uptake of ionizable substances by diffusion is highly dependent on the pH of the medium in which they occur. If the pH of the medium suppresses their ionization, uptake is greatly enhanced. Further, if there is a pH gradient across the cell or tissue membranes, such that a compound is more ionized on one side than on the other, the compound will move from the medium of lower ionization to the medium of greater ionization. Substances which do not ionize, e.g. hydrocarbons, may accumulate in the lipid phase of cell membranes or in fat deposits.

2.1.2 Catalysed diffusion

Catalysed diffusion requires the presence in the cell membrane of a carrier which can combine with the toxicant and move freely across the membrane. Since this is still diffusion, the toxicant will be transported down the appropriate electrochemical gradient. In many cases of catalysed diffusion the carrier appears to be a protein, which may be inactivated by any chemical which attacks protein. Such carrier proteins have been called permeases because of their similarity to enzymes. One aspect of this is the specificity of their binding properties. The carriers will bind only to a limited number of substances of similar chemical structure. These substances can compete for the binding site or sites and one may inhibit the binding of another and, therefore, inhibit its uptake. This phenomenon is called competitive inhibition. Conceivably, a competitive inhibitor could be used to prevent uptake of a toxicant where its presence is suspected. Equally, a toxicant may compete with a nutrient for a permease binding site, inhibiting uptake of the nutrient and producing an enhanced toxicity. Like enzyme-substrate binding, permease-toxicant binding is dependent on temperature, pH and other physicochemical interactions.

Not all carriers involved in catalysed diffusion are permeases. Ions may be transported by relatively simple hydrocarbons called ionophores. Ionophores have been isolated from some micro-organisms and also produced synthetically. The kinetic properties of ionophores are very similar to those of permeases, i.e. a given ion may competitively inhibit transport of another which is similar in its physicochemical properties.

2.1.3 Active transport

Like catalysed diffusion, active transport requires the presence of cell membrane carriers which bind to the toxicant but the toxicant is moved against the electrochemical gradient. For this to happen energy must be expended. This energy appears to be supplied in most cases in the form of ATP (adenosine triphosphate). The net result is that substances absorbed by active transport are accumulated by cells, even from solutions containing very small concentrations.

All the active transport carriers are thought to be proteins and, like the catalysed diffusion carriers, are classified as permeases. However, the active transport permeases have the capacity to hydrolyse ATP. It is thought that the energy released by this hydrolysis causes a conformation change in the permease molecule, allowing it to transport attached substances against the electrochemical gradient. In the best studied example of active transport, the human erythrocyte sodium pump, it seems clear that the permease is a plug of protein penetrating through the membrane.

The specificity of the enzyme-like binding site ensures that only chemically similar substances can be transported. Such substances may also compete for the site. This may be particularly important in relation to active transport systems which have evolved to ensure adequate uptake of essential nutrients. Uptake of non-essential substances by these systems may lead to harm from nutrient deficiency.

2.1.4   Uptake from the gastro-intestinal tract

For any given compound, the likelihood and the consequences of absorption will vary from one part of the gastro-intestinal tract to another (Fig. 2.1). For example, toxicants absorbed from the mouth will not have been modified by the gastric juices, nor will they be immediately exposed to possible detoxification by the liver (see Section 2.4). Thus, absorption from the mouth is often associated with a rapid and prolonged toxic response. Absorption from the stomach is characteristic of strongly acidic compounds, un-ionized at the low pH of the stomach contents, caused by hydrochloric acid secretion. Many weakly acidic, neutral and basic compounds are absorbed from the small intestine where the acid from the stomach becomes neutralized. Pinocytosis may contribute to the transfer of substances across the intestinal wall. Some toxicants may be absorbed from the colon.

Although most toxicants absorbed from the gastro-intestinal tract pass into capillaries and then to the hepatic portal vein and the liver, only part of the fat absorbed takes this route. Much of it passes into the lymphatic vessels, called lacteals, as an emulsion. This emulsion passes to the thoracic duct and enters the jugular vein. Thus, lipid toxicants may bypass the liver and avoid detoxification there.

Alimentary motility may influence gastro-intestinal absorption. Accelerated stomach emptying reduces the probability of absorption from the stomach. Increased intestinal peristalsis decreases intestinal absorption by speeding up removal of food from the intestine but simultaneously promotes absorption by improving mixing of the intestinal contents. The net effect reflects the balance between the two processes. Another significant factor is gastro-intestinal secretion. Reference has already been made to the acid secretion in the stomach. Secretion of mucus throughout the gastro-intestinal tract tends to inhibit absorption. Secreted hydrolytic enzymes (Table 2.1) may either destroy toxic compounds or even increase their toxicity, sometimes by

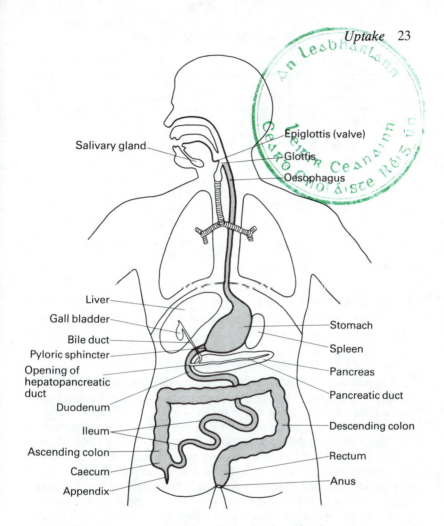

Fig. 2.1 Diagram of the human alimentary canal and associated organs.

promoting enterohepatic circulation (see Section 2.2.3). The intestinal blood flow may also affect absorption. In general, the greater the intestinal blood flow, the greater the intestinal absorption.

Finally, there is the nature of any foodstuff carrying the toxicant to be considered. Particle size will determine the rate of solution of a sparingly soluble substance, i.e. the larger the particle size the slower the rate of solution. Chemical interaction of a toxicant with other components of the gastro-intestinal contents may alter its availability. For example, complexing agents may bind toxic

Table 2.1   Principal hydrolytic enzymes of the mammalian gastro-intestinal tract.

Source	Part of the tract	Enzyme	Compounds hydrolysed
Salivary glands	Mouth	α-Amylase (ptyalin)	Starch, α-1,4-glucosides
Gastric glands	Stomach	Pepsin	Proteins, peptides
Pancreas	Small intestine	Trypsin	Proteins, peptides
		Chymotrypsin	Proteins, peptides
		Carboxypeptidase	Proteins, peptides
		Elastase	Proteins, peptides
		DNAse	DNA
		RNAse	RNA
		Lipase	Glycerol esters of fatty acids
		Phosphatase	Phosphate esters

metals, either preventing or facilitating their absorption. On the other hand, metals may precipitate some toxic substances so that they cannot be absorbed.

2.1.5   *Uptake from the skin or respiratory system*

Apart from the gastro-intestinal tract other possible sites of absorption of toxicants are the skin, the trachea and the lungs or, in aquatic animals, the gills. In intact skin the main barrier to absorption is the epidermis. This is composed mainly of lipoprotein and only lipid soluble compounds can pass through it easily. Ions and lipid insoluble molecules mostly penetrate the skin very slowly by passage through the hair follicles and sebaceous glands.

Uptake from the lungs is particularly important to terrestrial animals because this is the main route for atmospheric toxicants, either as gases or particulates (see Chapter 7). Gases which are very water soluble do not reach the alveoli because they dissolve in the water of the mucous membranes with which they first come in contact. Lipid soluble gases which reach the alveoli may be readily absorbed into the blood. Particles penetrate the lungs in inverse proportion to their size. Those less than 500 nm in diameter may enter the alveoli and remain there for very long periods. Such particles pose the greatest risks (see Section 7.4.2).

The gills of aquatic animals are specialized organs for gas exchange and are highly vascular. Further, various systems have been evolved by different organisms to ensure that the water in contact with the gills is continuously changed with that of the surrounding environment. Thus, there is a high likelihood that water-borne toxicants will be absorbed by the gills before they are taken up in any other way. Again, lipid soluble substances are taken up preferentially but the presence of detergents in water may break down the natural lipoprotein barriers to uptake of water-soluble materials making them much more toxic than they would otherwise have been.

2.1.6   Uptake by tissues

Once absorbed, toxic substances are transferred to the tissues from the general circulation. Fundamentally, the processes involved are much the same as already discussed in reference to uptake from the gastro-intestinal tract. Again, diffusion across the lipoprotein cell membranes is the prime means of transference and so lipid soluble compounds tend to be transferred preferentially. Traditionally, physiologists have defined three barriers to free transfer of metabolites in the mammalian body: the blood-brain barrier; the blood-cerebrospinal fluid barrier; and the placental barrier. These do not seem to bar transfer of lipid soluble compounds. The selectivity of these barriers seems to apply mainly to polar compounds. In some cases, the selectivity may be less in preventing uptake then in active excretion. For example, some organic anions and quaternary ammonium cations are actively removed from the cerebrospinal fluid by epithelial cells. The placental barrier should probably be distinguished from the other two, which are difficult to define anatomically, because it is a clearly-defined barrier of actively metabolizing tissues, the function of which is to separate the maternal and foetal circulations. The active metabolism of the placenta may alter the chemical nature of toxic substances passing through it and this must be allowed for in any consideration of possible effects of toxicants on the foetus.

2.2   Excretion

2.2.1   General considerations

Once a substance has been absorbed, its harmful effects will be minimized if it is rapidly excreted. Excretion occurs mainly in the urine or bile but may also occur in expired air, sweat, milk, saliva, gastro-intestinal secretions, genital secretions and by normal turnover of hair and skin.

2.2.2   Excretion in urine

Excretion in the urine depends upon the properties of the kidney (Fig. 2.2). Glomerular filtration produces an ultrafiltrate of blood plasma containing toxic substances and their derivatives in roughly the same concentrations as in the blood. Binding of toxicants to plasma proteins may prevent the toxicants appearing in the ultrafiltrate. The ultrafiltrate enters the renal tubules where it is modified both by reabsorption of components into the blood and by secretion of others from the blood. These two processes follow the same principles as outlined for gastro-intestinal absorption.

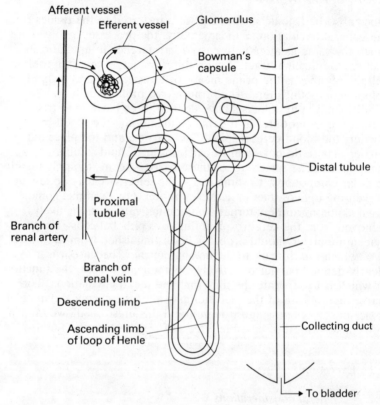

Fig. 2.2   Diagram of the fundamental unit of the mammalian kidney—the nephron.

Lipid soluble un-ionized compounds move down concentration gradients by diffusion. Ionizable compounds move from the medium in which they are least ionized to the medium in which they are most ionized. Thus, whether an ionizable compound is reabsorbed from the glomerular filtrate in the renal tubules, or secreted into it, depends on the pH of the filtrate since the blood pH stays fairly constant. In addition to these passive processes involving no input of metabolic energy, the renal tubular epithelium has active transport systems. One of these has the capacity to secrete strong organic acids and bases into the filtrate as it is transformed into urine. This system can operate against considerable concentration gradients. Since it alters the pH of the filtrate it has a profound effect on the diffusion of ionizable compounds across the membranes of adjacent cells. Other active transport systems have evolved in the renal tubule epithelium to reabsorb essential metabolites such as amino acids and glucose. Toxic substances similar to these metabolites will compete for uptake. Not only will such compounds accumulate in the organism but part of their toxicity may be due to their causing loss of the corresponding metabolites in the urine.

2.2.3 Excretion in bile

Excretion in bile is a more effective protection against toxicity than renal excretion since it involves much less exposure of tissues to the toxicant. Blood, carrying compounds and salts absorbed from the gastro-intestinal tract, enters the liver by the hepatic portal vein and passes into the hepatic sinusoids. From the sinusoids many substances are absorbed into the hepatic parenchymal cells. There, metabolic transformations, described in Section 2.4, may occur. The absorbed substances, or derivatives, may eventually leave the parenchymal cells by one of two routes, either back into the sinusoids or into the bile. For some highly polar compounds, e.g. the bile salts, bilirubin glucuronide and most conjugates (see Section 2.4), excretion into the bile is an active process. If substances pass back into the sinusoids the blood will carry them into the hepatic vein for ultimate excretion by the kidneys. If substances enter the bile duct the bile will carry them to the duodenum. If the toxicants and/or their derivatives are nonpolar there is a good chance that they will be reabsorbed from the intestine, reenter the liver and be reexcreted in the bile. This phenomenon is known as the enterohepatic circulation and can lead to cumulative poisoning of liver cells. Polar conjugates of nonpolar compounds may be hydrolysed in the intestine, permitting the nonpolar compounds to be recirculated.

2.2.4 Excretion by other routes

Volatile compounds may be excreted in expired air. Lipid soluble compounds may be excreted in milk. Some toxicants may reenter the gut in gastro-intestinal secretions or from saliva. Genital secretions may contain toxic substances and these may inhibit fertilization.

Excretion in hair and skin has been of particular interest in relation to arsenic and mercury. Hair has attracted most attention because it is easy to obtain for analysis and because its chemical stability makes it possible to detect poisoning long after the event, even after the person or animal poisoned has died and other tissues have completely decomposed. Also, the growth of hair means that analysis along the length of any single hair gives a time course of incorporation of the poison. Such analyses of hair have shown that Napoleon Buonaparte suffered arsenic poisoning and that the poet Robert Burns may well have been poisoned with mercury, probably prescribed by his medical practitioner.

2.3   Chemical Localization and its Consequences

2.3.1   Lipid solubility

Toxic substances which remain in the body show localization in tissues, cells and cell components, depending upon their properties. Substances of high lipid solubility become localized in adipose tissues. Compounds of this kind include chlorinated hydrocarbons, e.g. DDT, DDE, the epoxide derivatives of aldrin, and methyl mercury. Frequently, lipid solubility is associated with chemical stability and, therefore, these compounds accumulate, potentially to dangerous levels.

2.3.2   Protein binding

Protein binding effects distribution of toxic substances within the body. An important factor here is binding to plasma proteins, especially serum albumin. Serum albumin normally picks up metabolites such as keto-acids, fatty acids, bilirubin and hormones. Displacement of these by a competing compound may be sufficient to cause toxic effects. For instance, sulphonamides displace bilirubin causing hyperbilirubinaemia which may lead to kernicterus (bilirubin-induced brain damage).

Nonantigenic toxic substances may bind to serum globulins to produce antigens, or form stable complexes with serum albumin which have the same effect (Fig. 2.3). Nonantigenic molecules of this kind are referred to as haptens. The derived antigens induce sensitivity in lymphoid cells which may be sufficient to make the cells subsequently respond to the hapten on its own. Ill effects are not usually observed on the first exposure to such a hapten, but subsequent exposure may cause reactions of varying degrees of severity, even death. This phenomenon is known as allergy or hypersensitivity. Hypersensitivity may be delayed or immediate. In delayed (cellular) hypersensitivity, lymphoid cells react with the antigen or hapten in any organs where they meet and gradual damage to these organs results. In immediate (humoral) hypersensitivity, circulating antibodies react with the antigen or hapten and within minutes tissues throughout the body may be affected, particularly blood vessels, smooth muscle and connective tissue. Delayed hypersensitivity may result, for example, from skin exposure to penicillin or to the toxic compounds produced by poison ivy. Immediate hypersensitivity is a consequence of the

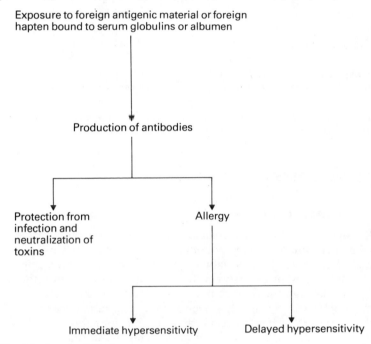

Exposure to foreign antigenic material or foreign hapten bound to serum globulins or albumen

Production of antibodies

Protection from infection and neutralization of toxins

Allergy

Immediate hypersensitivity

Delayed hypersensitivity

Fig. 2.3   Immunological responses.

release of substances such as histamine and serotonin following the antigen-antibody reaction at the surface of mast or basophilic cells. Histamine and serotonin act on blood vessels and smooth muscle. In the skin, this reaction causes dilation of the blood vessels and localized swelling. In the lungs, the reaction causes contraction of the smooth muscle which may be fatal. Other substances involved in these reactions are bradykinin and SRS (slow-reactive substance). Sometimes interaction of antigen with antibody in blood vessels leads to injury to the vessels and formation of a clot. This is called the Arthus reaction. Small antigen-antibody complexes may penetrate the blood vessel walls in certain organs, especially the kidney, and lodge in the basement membrane where they act as an irritant. This may lead to glomerulonephritis.

2.3.3   Nucleic acid binding

Binding of toxic substances to nucleic acids may not play much part in selective tissue localization, but will lead to cell localization in ribosomes and nuclei. This is of particular concern because interaction with nucleic acids is frequently the basis for mutagenicity and carcinogenicity. Amongst the known carcinogens which interact with nucleic acids are polycyclic hydrocarbons; polycyclic amines, after conversion to hydroxylamino derivatives and *o*-amino phenols; aminoazobenzene derivatives; aliphatic dialkyl nitrosamines, after conversion to mono-alkylnitrosamines which decompose spontaneously to diazoalkanes which alkylate nucleic acids and proteins irreversibly; other aliphatic alkylating agents such as epoxides, lactones, ethylene imines and nitrogen and sulphur mustards.

2.4   Hepatic Metabolism

2.4.1   General considerations

Although only unmodified toxic substances in the organism have so far been considered, many potentially toxic substances are metabolized by cells, especially by the hepatic parenchyma cells. Since most substances absorbed from the intestine pass directly to the liver through the hepatic portal vein, metabolism by the hepatic parenchyma cells has been regarded as an important defence system against toxicants and the transformations involved have been referred to as detoxification. This is unfortunate, because, while these transformations may decrease toxicity in some cases, they may increase it in others. Thus, few generalizations are possible as to the ultimate consequences of hepatic metabolism, although a great deal has been discovered about the relevant biochemistry and physiology.

2.4.2   Microsomal transformations – cytochrome P_{450}

The metabolic transformations of toxic substances in the liver are mainly associated with microsomal enzymes, i.e. with enzymes occurring in the endoplasmic reticulum, since microsomes are vesicles formed from the endoplasmic reticulum during cell homogenization. The reactions catalysed by microsomal enzymes may be categorized as follows: oxidation, reduction, hydrolysis, dealkylation, deamination, desulphuration, dehalogenation, ring formation, ring scission and conjugation (see Table 2.2). Conjugation of organic compounds with sugars, amino acids or their derivatives gives a product which is usually more water soluble than the original organic compound and which is, therefore, more easily excreted. Almost all of the reactions mentioned are associated with the cytochrome P_{450} system, also sometimes referred to as 'mixed function oxidase' or 'microsomal drug metabolizing enzyme'. Exceptions are hydrolysis and conjugation, though the cytochrome P_{450} system is involved in conjugating compounds with cysteine to form mercapturic acids.

The cytochrome P_{450} system acts as a mono-oxygenase, splitting molecular oxygen into two atoms, one of which enters the enzyme

Table 2.2 Liver microsomal transformations.

Reaction	Description
Oxidation	Conversion of carbon atoms to hydroxymethyl, aldehyde or carboxylic acid groups; conversion of sulphite to sulphate
Reduction	Removal of oxygen or addition of hydrogen
Hydrolysis	Cleavage of ester, amide and other bonds by addition of water
Dealkylation	Removal of alkyl groups, e.g. methyl or ethyl groups, attached to nitrogen, oxygen or sulphur atoms
Deamination	Removal of amino groups
Dehalogenation	Removal of chlorine and other halogens
Ring formation	Formation of cyclic molecules from straight chain compounds
Ring scission	Cleavage of cyclic molecules
Conjugation	Covalent linkage to another molecule

substrate, while the other participates in the formation of water. Apart from cytochrome P_{450}, the system has several components including nicotinamide adenine dinucleotide phosphate (NADP) a flavoprotein and a non-haem iron protein. The system is found in both vertebrates and invertebrates. There is some suggestion that the activity of the system may be lower in aquatic vertebrates than in terrestrial vertebrates. Perhaps one of the most interesting properties of the system is that it can be induced by exposure of organisms to various compounds, especially those with a high lipid solubility and a slow rate of metabolism. Thus in cases where the system detoxifies these compounds, exposure of animals to subtoxic levels over a period of time will generate tolerance to much higher levels. With regard to compounds which are converted to more toxic derivatives, similar exposure of animals to subtoxic levels will increase sensitivity of response. It is even possible that exposure to a nontoxic substance, which induces the cytochrome P_{450} system, may make an animal either more resistant or more sensitive to a totally different substance which is toxic.

Other changes may accompany increased activity of the cytochrome P_{450} system. These include accelerated steroid synthesis, modification and deactivation, increased ascorbic acid biosynthesis and increased porphyrin biosynthesis. Unlike the other effects, increased porphyrin biosynthesis is not associated with microsomal enzymes but is caused by induction of the mitochondrial enzyme, delta-aminolaevulinic acid synthetase, which is the first enzyme in the porphyrin biosynthesis pathway. Excessive production of porphyrins in human beings produces symptoms of the illness called porphyria. This illness is normally described as being due to an inborn error of metabolism since it can be related to a recessive gene in the human population. Clearly, sufferers from genetic porphyria will be unusually susceptible to the harmful effects of compounds that have the inductive effect described. The symptoms of porphyria range from skin photosensitization and dermatitis to chronic mental aberrations. As well as substances which induce the cytochrome P_{450} system, there are some known to inhibit it. An example of this kind is SKF 525A (β-diethylaminoethyl-diphenyl-propyl acetate). This compound is used to prolong the activity of drugs that are deactivated by the cytochrome P_{450} system. Possibly, it might also be used to inhibit the formation of toxic derivatives of other compounds.

As well as the immediate biochemistry of the cytochrome P_{450} system, the biological status of the organism must be taken into account in assessing its ability to metabolize toxic substances. Genetic factors determine whether the metabolizing enzymes can be present or not and, perhaps, also the number of enzyme molecules that can be synthesized under given conditions. For example, several genetic disorders of humans are associated with reduced glucuronyl transferase activity, and hence with poor conjugation of compounds to glucuronic acid. This reaction is essential for effective excretion of bilirubin, the chief breakdown product from porphyrins. Hence, exposure of persons with reduced glucuronyl transferase activity to compounds which compete with bilirubin for the enzyme can result in elevated blood levels of bilirubin which in turn may cause brain damage (kernicterus).

2.4.3 Genetic, physiological and environmental factors controlling hepatic metabolism

While genetic factors determine the potential of an organism for enzyme synthesis, physiological and environmental factors determine when and to what extent this potential is realized. The developmental stage which an organism has reached can make an appreciable difference to its metabolic capacity. In mammals, from birth onwards there is a marked increase in the activity of many liver enzymes. Depending upon the enzyme, the increase to maximum levels may occur within a few hours or over a period of up to two months. Newborn mice, rats, guinea pigs and rabbits have been found to lack cytochrome P_{450} and to have little capacity for production of glucuronides and glycine conjugates. Thus, the newborn animals are highly susceptible to compounds which adults can detoxify by these systems. Sex differences are superimposed on the effects of development. Adult male mammals frequently, but not always, metabolize toxic substances faster than females. This may be beneficial or it may not, depending on the toxicity of the products. Hormonal status of the organism is another factor to be taken into account, but it is difficult to isolate effects of individual hormones and there are few generalizations that can be made. During late pregnancy, the glucuronide conjugation of foreign compounds is markedly reduced, probably owing to the high circulating levels of progesterone and pregnanediol. Sulphate conjugation, demethylation and hydroxylation may also be inhibited. Oral contraceptives based on progesterone-like compounds have similar effects and this should be borne in mind when drugs are being prescribed for women taking these contraceptives. Like the effects of hormones, the effects of starvation on metabolism of toxic substances is highly variable, as reported in the literature, but it seems that deficiencies of protein, ascorbic acid, calcium or magnesium can impair the activity of microsomal enzymes. Any disease which damages the liver has a similar effect.

All the preceding physiological effects will inevitably incorporate an environmental dimension. Environmental factors which may warrant special attention include: stress, which tends to increase metabolism following the pituitary-adrenal response; ionizing radiation, which may impair microsomal enzyme activity; and exposure to compounds other than the toxicant of interest, since this may induce microsomal enzyme systems.

2.5 Synergistic and Antagonistic Effects

Possible interactions between different substances, both toxic and nontoxic, have been described. Where the effect of a toxic substance is enhanced by exposure of the affected organism to another substance, synergism is said to occur between the two substances and the latter substance, whether toxic itself or not, is said to be synergistic with the former. Where the effect of a toxic substance is reduced by exposure of the affected organism to another substance, antagonism is said to occur and the latter substance is said to be antagonistic to the former. These terms apply whether the animal is exposed to the substances simultaneously or with a time differential. Sometimes, varying the time interval between exposure to the toxic substance and to the secondary substance can produce results varying from synergism to antagonism or vice versa. Synergism and antagonism can result from effects on the microsomal enzymes; from competition for binding sites, displacing the toxic substance from an inactive or active complex; or from an effect on uptake or excretion, altering the net accumulation of the toxic substance within the organism. There are many other possibilities.

2.6   Variations in Metabolism among Animals – Comparative Toxicology

Finally, the point must be made that no two organisms are exactly the same. The ability to metabolize toxic substances varies from individual to individual, species to species, genus to genus, etc. It is this variability which makes possible the development of selective pesticides. However, most of the metabolic processes outlined have been detected in all the major groups of animals studied. A few qualitative differences have been reported. Where Mammals conjugate compounds with glucuronic acid, insects and molluscs use glucose. One mammal, the cat, forms glucuronides with difficulty and, therefore, is unusually susceptible to certain toxic substances, notably phenols. While terrestrial animals, fish and crustacea can conjugate some foreign compounds with glycine, they may also use other amino acids for this purpose. For example, man and the primates use glutamine, birds and reptiles use ornithine, and arthropods use arginine and glutamine. Of all the species examined, reptiles are the only group incapable of acetylating foreign compounds. Insects appear to have a unique method of eliminating metals by converting them to insoluble sulphides.

  At the time of writing, the preceding statements seem to be reasonable generalizations but the field of study which they represent, comparative toxicology, is so all-embracing that it would be surprising if there are no exceptions. Further, these statements are largely qualitative in nature. Many quantitive differences remain to be defined.

3 Metabolism of Toxic Substances by Plants

Plant toxicology has been restricted largely to studies of responses to pathogen toxins, herbicides and air pollutants. Plants in a poor physiological state usually show increased sensitivity to toxic substances. Anything which drastically upsets plant cell metabolism tends to trigger the production of aromatic compounds such as mono- and dihydric phenols, phenolic glycosides, flavonoids, anthocyanins, aromatic amino acids and coumarin derivatives. These compounds protect the plant against bacterial or fungal infection. Increased activity of several oxidative enzymes including cytochrome oxidase, phenol oxidase, ascorbic acid oxidase and peroxidase has been reported to correlate with resistance to pathogen toxins. Exposure of plant cells to toxicants, or even to ions of some heavy metals, may induce the formation of phytoalexins. Phytoalexins are substances which are frequently produced by plant cells following infection and which protect against further infection by a non-specific antibiotic action. They vary chemically but are often phenolic with multiple six-membered ring structures (Fig. 3.1). Finally, lesions in metabolism of cells in certain parts of the plant may stimulate translocation of materials to these cells. Presumably, the translocated substances will sometimes alleviate adverse effects.

3.1 Uptake

Before the responses described above can be observed any toxic substance involved must have reached the cells affected and altered their metabolism. If the toxic substance is released by a pathogenic fungus or bacterium, it will act mainly on cells in the immediate vicinity of the infection. If the toxic substance is an environmental contaminant, it must have been absorbed by the plant. The two most likely sites of absorption are the roots and the leaves. Absorption by the roots is largely from the soil solution and chiefly involves inorganic ions and water soluble compounds. Such absorption is highly dependent on the nature of the soil solution, which is affected by the affinity of soil particles for solutes (see

Structure	Trivial name	Source

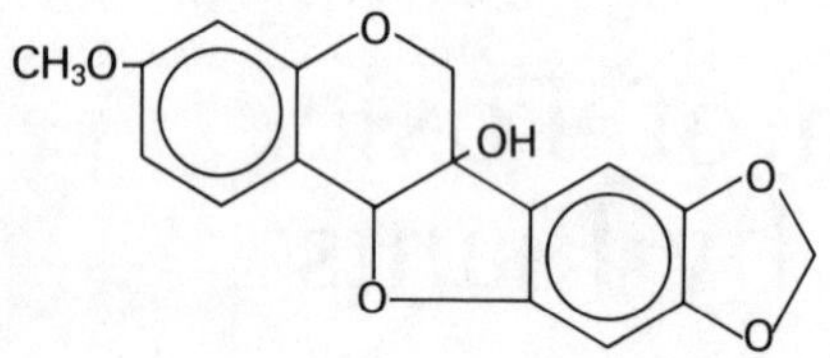

Pisatin — *Pisum sativum* (pea)

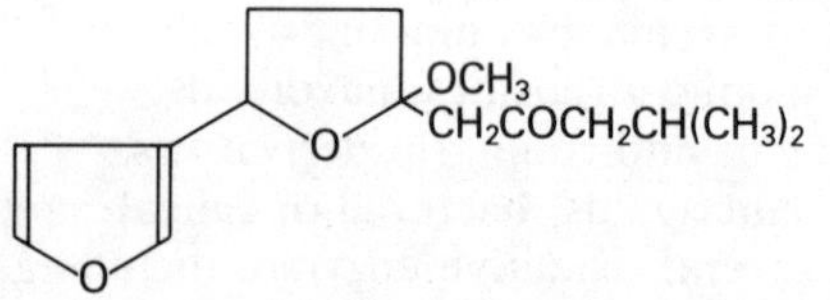

CH₃O (structure) — OH

Ipomeamarone — *Ipomaea batatis* (sweet potato)

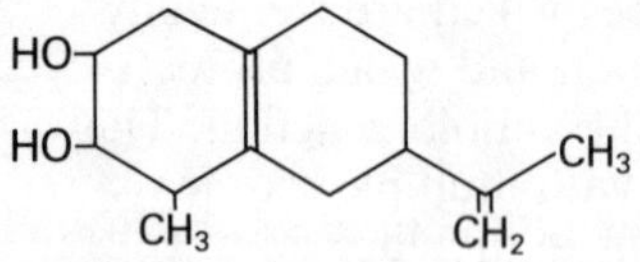

OCH₃ CH₂COCH₂CH(CH₃)₂ (structure)

Rishitin — *Solanum tuberosum* (potato)

HO, HO, CH₃, CH₂, CH₃ (structure)

HOCH₂CHOHCH = CH (C≡C)₃CH=CHCH₃   Safynol   *Carthamus tinctoria*

Fig. 3.1 Structures of some phytoalexins from higher plants.

Section 9.1) and by the activity of soil micro-organisms which metabolize these solutes (see Section 4.5). Otherwise, root cells and all other plant cells show absorption properties similar to animal cells (see Sections 2.1.1–2.1.3). Thus, uptake of substances is mainly by passive diffusion, favouring lipid soluble molecules. For some substances uptake is active, involving the expenditure of metabolic energy, and these and related compounds can be concentrated against large electrochemical gradients. Toxicants absorbed by roots are frequently retained in root storage organs, and tend to persist there much longer than they would in stems or leaves. Absorption by leaves can be either of gaseous substances from the atmosphere, or of substances in solution in rain, snow, or sprays applied for pest control. Absorption by leaves is most marked for lipid soluble compounds since these can penetrate the waxy cuticle. Gaseous substances may enter through the stomata but again lipid solubility contributes to their uptake through the leaf cell membranes.

The exposure of any particular leaf to toxicant depends upon the morphology of the plant. Density of foliage may prevent toxic substances reaching inner leaves. Narrow, waxy, or sculptured leaves retain less deposited material than broad, hairy leaves. Retention of aqueous droplets depends upon wettability. Within a given species, immature leaves tend to be less wettable than mature ones; lower surfaces of leaves are more difficult to wet than upper surfaces; monocotyledonous plants wet less easily than dicotyledons etc. Toxicants can also penetrate the seed coats of many seeds, depending upon the same processes already outlined for roots and leaves. Penetration decreases markedly as the seeds reach maturity. This may be an important consideration in the application of pesticides to plants during seed development and in their use as seed dressings.

Rates of uptake of toxicants by plants are highly dependent on environmental factors such as day-length; light quality and intensity; temperature; moisture content and chemical composition of soils; and air humidity.

3.2 Translocation

Once absorbed, substances may pass from cell to adjacent cell through cytoplasmic strands called plasmadesmata which traverse the surrounding cell walls. Alternatively, the absorbed substances may pass into the aqueous solution in the cell walls and diffuse through it. Such diffusion is restricted in the root by the Casparian strip, a suberized part of the endodermal cell wall, which separates the cortex from the vascular tissue. Once the substance enters the vascular tissue, it may be transported rapidly from the roots to the leaves via the transpiration stream in the xylem, or more slowly from the leaves in either an upward or downward direction in the phloem. Phloem transport is complex and not fully understood but there seems little doubt that toxic substances entering the phloem could have a very serious adverse effect on the plant. Absorption of substances from the vascular system varies throughout the plant depending upon the nature of the tissues, their hormonal status and environmental conditions. Hormonal status and the correlated developmental stage may be particularly important. Apical meristems and regions with high concentrations of growth hormones tend to accumulate nutrients and related substances, often at the expense of ageing or senescent tissue.

3.3   Metabolism and Excretion

Toxic substances which enter plant cells may react with cell components to exert harmful effects or they may be metabolized directly. Metabolism of toxicants appears to be similar to that in animal cells, by transformations involving hydrolysis, conjugation and the action of the cytochrome P_{450} mixed function oxygenase system. In addition, plant cells may sequester toxic substances in the vacuole. Apart from the shedding of dead tissues which have been poisoned and, more rarely, loss by evaporation, guttation or root exudation, this is the principal means of excretion of toxicants and their derivatives.

3.4   Pathogen Toxins

The toxins produced by pathogens are mostly ill-defined, both chemically and with regard to their fundamental effects on plant cells. Some appear to be proteolipids and one of the most studied, that from *Helminthosporium sacchari* which causes eyespot disease in sugar cane, has been identified as a sugar, helminthosporoside. Some, e.g. helminthosporoside, are highly selective in their toxicity and this suggests that they must react with specific receptor molecules, probably proteins. The initial consequences of this reaction are often changes in membrane permeability, though direct effects on mitochondria have also been reported.

3.5   Herbicides
See also Section 5.5.

3.5.1   General considerations
Much more is known about the effects of manmade herbicides than about the effects of naturally occurring toxic substances. At present the herbicides can be divided into five groups: photosynthetic inhibitors; photosynthetic energy deviators; inhibitors of chloroplast development and carotenoid formation without any prior effect on photosynthesis; uncouplers of oxidation phosphorylation; growth regulators. In the following sections herbicides will be referred to by their trivial names. Full chemical names are listed in Appendix 4.

3.5.2 Photosynthetic inhibitors

About 45% of all herbicides can be classified as photosynthetic inhibitors. These include the substituted ureas, e.g. monuron and diuron; the triazines, e.g. atrazine, simazine and simeton; the hydroxybenzonitriles, e.g. ioxynil and bromoxynil; the uracils e.g. bromacil and isocil; the amide propanil; and the carbamates, e.g. propham and chlorpropham. It is thought that many of these compounds act by preventing electron transfer between photosystem II and photosystem I. Response to these herbicides is fairly slow.

3.5.3 Photosynthetic energy deviators

Herbicides classified as photosynthetic energy deviators belong to the bipyridylium group, e.g. paraquat and diquat. These herbicides compete for electron flow at the reducing end of photosystem I, accepting electrons that would otherwise have been passed on to ferredoxin. The phytotoxic effect of these herbicides is rapid in the presence of oxygen to reoxidize the reduced bipyridylium ion.

3.5.4 Inhibitors of chloroplast development

Amitrole, dichlormate, pyrichlor, haloxydine and Sandoz 6706 are classified as inhibitors of chloroplast development. They do not block photosynthesis in fully-developed chloroplasts. Their fundamental action is to inhibit the formation of internal structrue and block carotenoid synthesis.

3.5.5 Uncouplers of respiration

The herbicides that act primarily as uncouplers of respiration are mainly related to 2,4-dinitrophenol (DNP). They include dinoseb, DNOC and PCP. The hydroxybenzonitriles, bromoxynil and ioxynil, and benzimidazoles also uncouple oxidative phosphorylation and there is a good correlation between herbicidal efficiency and uncoupling power. However, to call them herbicides gives a false impression of their specificity. Since production of adenosine triphosphate (ATP) by oxidative phosphorylation is the main source of the energy essential for the maintenance of life in all aerobic organisms, these compounds can kill most living things. They are general biocides and human fatalities have been caused by carelessness in their application.

3.5.6 Growth regulators

Finally, there are the herbicides that act as growth regulators. Of these, the best known are 2,4-D, 2,4,5-T (see Section 5.5) and

MCPA. Treatment of plants with these compounds commonly leads to stem elongation and curvature, production of secondary roots and other types of abnormal growth. These superficial phenomena reflect suppression of cell division and elongation in the apical meristem, as well as abnormal cell division in the vascular tissue.

One of the newer herbicides, glyphosate, may also be classified as a growth regulator, though it is more properly thought of as a growth inhibitor. It appears that it causes plant death by inhibiting the synthesis of aromatic amino acids and hence of protein.

3.6 Atmospheric Toxicants

Atmospheric toxicants such as ozone, peroxyacyl nitrates, sulphur oxides and nitrogen oxides are discussed in Chapter 7, but their effects on plants will be considered here in a more general context. The effects so far described for all these substances are remarkably alike. Acute injury with rapid tissue death is characterized by collapsed marginal or intercostal leaf areas, which dry and bleach or become brownish red. Severe chronic injury produces leaf yellowing until most of the chlorophyll and carotenoids are destroyed and interveinal segments of the leaf are white. Histologically, these changes are associated with plasmolysis, granulation and disorganization of affected cells, sometimes with abnormal production of pigments. Less severe or sublethal chronic injury is, as with animals, difficult to ascertain and has been referred to as hidden, invisible or physiological injury. Some of the effects included in this category are reduced water uptake, stunted growth, stomatal closure and reduced photosynthetic efficiency. Growth inhibition is particularly noticeable with sulphur oxides (see Section 7.3.2).

3.7 Environmental Factors

The nature and degree of the response of a plant to any toxic substance depends upon the environmental conditions. Light, water, temperature and nutrition are perhaps the main factors to be considered. Some conditions will be antagonistic to the toxic substance and others will be synergistic. Synergism is almost certain where other harmful substances are present. As was noted with animals, there is a need for the investigation of such interactions, particularly where a number of toxic substances are present at levels which individually do not produce perceptible effects.

3.8 Lichens and Bryophytes

The preceding considerations relate primarily to angiosperms but special attention must be paid to lichens (Fig. 3.2.) and bryophytes (liverworts and mosses), since these plants show an extremely high sensitivity to air pollutants and have been suggested to be biological pollution indicators. Because different species of lichens and bryophytes have different sensitivities to atmospheric toxicants, analysis of the species in a given area can give a measure of the mean level of pollution once the sensitivity has been defined for areas of that type. Transplants of lichens and bryophytes from unpolluted areas may be used to verify observations of the natural community.

Fig. 3.2 Healthy growth of lichens on a tree stump in Vermont, USA, shows absence of atmospheric toxicants.

The fundamental effects of atmospheric toxicants on lichens and bryophytes seem to be chlorophyll degradation and chloroplast damage. This inhibits growth. Reproduction is also affected in many species, decreasing with increasing toxicant level. Toxicant sensitivity of any species of lichen or bryophyte is related to the buffer capacity of its substrate. For example, species occupying acidic substrates are usually more sensitive to sulphur dioxide than those on basic substrates. Possibly because of this, lichens growing on the bark of trees are more sensitive than those growing on stones which, in turn, are more sensitive than those growing on the soil. Thus, plants growing on trees have attracted the most attention in relation to pollution effects. Of these plants, trunk species are more sensitive than those on tree bases. Trunk species on nutrient-poor oak or ash bark are more sensitive than species on nutrient-enriched (eutrophiated) elm bark. This seems to be associated with the air particulates (see Chapter 7), especially those containing alkaline earth metals, which are trapped in the crevices of nutrient enriched barks and which increase their buffering capacity.

The resistance of species growing on tree bases, soil or stones to atmospheric toxicants is not related only to the basicity of the substrates. The shelter provided by surrounding ground vegetation reduces contact with polluted air and, during the winter when toxicant concentrations tend to be high, snow cover may provide further protection.

Inherent properties of the lichens and bryophytes which effect their resistance to toxicants include their cytoplasmic pH and their moisture level. It has been suggested that the marked resistance of the common grey-green encrusting lichen, *Lecanora conizaeoides*, is due to the nonwetting properties of its crustose thallus.

3.9   Plants as Pollution Indicators

Although cryptogamic epiphytes have been recommended as biological pollution indicators because they are easy to handle and show a wide range of sensitivity to toxicants, especially air pollutants, the possible use of other plants for this purpose should not be ignored. Plants have a great advantage over animals for this purpose in that, in the vegetative stage, they cannot move away from adverse conditions. It is certain that the sensitivities of plants to toxic substances will vary both qualitatively and quantitatively. Thus, analysis of vegetation in any area should give a measure of toxicant stress and may indicate what toxicants are involved. This is clearly one of the major areas for future research in environmental toxicology.

3.10 Plants as Toxicant Sources

Poisonous plants have been identified as a major cause of livestock loss in farming. It is difficult to obtain accurate figures but one recent estimate put losses in Texas alone at between 50 and 100 million dollars annually. Kingsbury has reported over 1000 species of plants in the USA and Canada that are toxic to livestock and it is likely that similar figures could be produced for other parts of the world.

Perhaps the best known of all poisons is hydrogen cyanide and many plants are known to be capable of releasing this substance when crushed. The hydrogen cyanide does not occur free within the plant but is released by hydrolysis of more complex compounds. Amongst the plants which can poison animals by acting as a source of hydrogen cyanide are young sorghum, arrow grass, certain species of acacia, cassava and white clover. The effect of hydrogen cyanide is to block oxidative production of energy by inhibiting the terminal respiratory enzymes.

One of the most common plant toxicants is oxalic acid. Most oxalate producing plants belong to the genus Rumex or genera of the Chenopodiaceae and Oxalidaceae families. Unfortunately, these plants are palatable to livestock and mass mortality can result from grazing pastures where they are present. One such plant, halogeton, has been blamed for incidents in western USA leading to the deaths of between 100 and 1200 sheep at a time. The toxic effects of oxalic acid are largely due to its forming insoluble salts with calcium and magnesium. These effects can sometimes be overcome by treatment with calcium borogluconate. The symptoms vary with the plant ingested. In sheep, excessive halogeton ingestion causes rapid, laboured respiration leading to coma and death. On the other hand, excessive intake of soursob, found in South Africa and Australia, causes hypocalcaemia and tetany. The latter are the most frequently observed consequences of oxalic acid poisoning but there is considerable variation, reflecting the many interactions that are possible.

Another group of plant toxicants that has attracted much attention comprises the organofluorine compounds, the first one to be reported being fluoroacetic acid. This compound occurs in a number of plants, especially *Gastrolobium bilobum* and *Oxylobium parviflorum* from Western Australia, the dry leaf of which may contain up to 1.25% fluoroacetic acid by weight. Other organofluorine compounds detected in plants include fluoro-oleic acid and fluoropalmitic acid. Poisoning with fluoroacetic acid in short-term dosing is cumulative and fatal at relatively small doses ($0.25-0.5$ mg/kg^{-1} for sheep) which vary considerably with different species. Death is caused by conversion of fluoroacetic acid to fluorocitric acid which inhibits oxidative production of energy which is essential to aerobic life.

Hydrogen cyanide, oxalic acid and fluoroacetic acid are relatively simple substances but, as research continues, more complex toxicants have been identified. The pyrrolizidine alkaloids, from various plants including *Crotal* spp. and *Trichodesma incanum*, cause liver, lung and kidney diseases in many animals. Sesquiterpenes from *Myoporum* spp. have been shown to cause extensive liver damage in livestock in Australia and New Zealand. Cardiac glycosides have been implicated in the toxicity of milkweeds to sheep in the western United States. Miserotoxin (3-nitro-l-propyl-β-D-glucopyranoside) has been identified as the toxic component of some *Astragalus* spp. which causes muscle weakness, loss of motor control and other nervous symptoms. Other *Astragalus* spp. accumulate selenium and convert it into highly toxic selenium-containing amino acids (see Section 6.15). Alkaloid teratogens have been isolated from Lupinus, Conium, Veratrum and related genera. Oestrogenic substances causing ewe infertility have been identified in certain cultivars of subterranean clover. The list is constantly being added to and the above are only an arbitrary selection of those that are already known.

A final point to be made is that toxic effects following ingestion by animals may be associated with plants which do not contain toxicants. This seeming paradox arises from the possibility of formation of toxicants from nontoxic precursors by gut micro-organisms. It has been suggested by Carlson and Dickinson (see Keeler, van Kampen and James, 1978) that acute bovine pulmonary edema (ABPE), a serious lung disease of cattle, is caused by 3-methylindole, generated in the rumen by fermentation of tryptophan and related compounds. The possibility of such occurrences serves to point out the problems of assessing toxicity and related hazards. It also brings out the general need for an increased study of comparative toxicology and of the specific need for more information on conversions of toxicants and toxicant precursors in the rumen.

4 Toxic Substances Released into the Environment by Micro-organisms

In 1888 Roux and Yersin reported that diphtheria bacillus produced a toxin or poison. Since then many other microbial toxins have been identified. They may be classified as endotoxins or exotoxins. The endotoxins are cell components, frequently part of the cell wall, which are released only when cells break down or lyse. They are usually antigenic complexes of protein, polysaccharide and lipid. The protein determines the antigenicity and the polysaccharide the immunological specificity. To some extent the lipid determines the toxicity. Therefore, endotoxins are not neutralized by homologous antibody since this binds to the largely nontoxic protein and polysaccharide. All endotoxins have similar pharmacological properties.

Exotoxins are distinguished from endotoxins by the fact that they are released from intact cells during the exponential or decline phase of growth. Most exotoxins are antigenic proteins which are neutralized by homologous antibody. Although exotoxins are much less stable to heat than endotoxins, their toxicity is much greater.

The bacterial endotoxins are nearly all produced by Gram-negative bacteria. Almost all of these toxins must be released within the circulatory system of animals in order to take effect. Therefore, their presence in the environment does not pose a hazard. The hazard comes from the risk of infection with the bacteria which produce the toxins.

Consideration of illnesses due to direct bacterial infection of organisms does not come within the scope of this book, and the reader who wishes to follow this further is referred to any good text on bacteriology. However, one bacterial endotoxin that should be mentioned is the toxin produced by *Clostridium botulinum*. The *C. botulinum* toxin was classed formerly as an exotoxin but it now seems established that it is released only from disrupted cells. This toxin has been associated with many cases of fatal food poisoning. A number of fungal and algal endotoxins have also been implicated in food poisoning.

Exotoxins are produced mostly by Gram-positive bacteria, with the exceptions of *Shigella dysenteriae* and *Vibrio cholerae* which are Gram-negative. Exotoxins are produced also by fungi and algae. Of the bacterial exotoxins, the enterotoxin produced by *Staphylococcus aureus* is of particular concern because it is one of the most common causes of food poisoning throughout the world. This toxin is referred to as an enterotoxin to distinguish it from other *S. aureus* toxins which are inactivated in the gut.

This chapter will examine the risks posed by *C. botulinum* and *S. aureus* toxins, and by toxins from selected fungi (mycotoxins) and algae. A final section will be devoted to toxic products of microbial metabolism which are serious environmental hazards, though they are not normally classified as toxins. These range from fairly complex derivatives of pesticides to simple compounds like ammonia.

4.1   *Clostridium botulinum* Toxins

4.1.1   *Botulism*

Botulism is the name given to the symptoms of poisoning caused by the *C. botulinum* toxins. Botulism affects many bird and animal species throughout the world and is frequently fatal. In man, symptoms include difficulty in swallowing, breathing and eye focusing, together with paralysis of the extremities. The only treatment is early administration of the type specific toxin antiserum. Since early diagnosis is difficult and typing of the toxin is time-consuming, it is rare for effective treatment to be possible.

4.1.2   *C. botulinum*

C. botulinum is an anaerobic, sporeforming, motile bacillus which is Gram-positive in the early stage of growth. Its natural habitat is soil, where it is found mostly as spores which can survive extreme environmental conditions, Six strains of *C. botulinum*, designated A, B, C, D, E and F, have been shown to be responsible for botulism. The toxins produced by these strains are given the same designations. There appears to be some selectivity with regard to the foodstuffs likely to carry the different strains. Type A is associated with home-canned vegetables and fruits, and with meat and fish. Type B is found mostly in meats, especially pork products. Type Cα occurs in fly larvae and rotting vegetation in alkaline ponds, while type Cβ is found in forage, carrion and pork

liver. Type D grows in carrion, type E in uncooked flesh from fish
and marine mammals and type F in home-made liver paste. Many
other foods may harbour the organism but this short list will give
some idea of the scope of the potential hazard.

Production of the toxins is associated with growth of
C. botulinum. This can occur under anaerobic conditions in most
foods with pHs above 4.5 (optimally between 5.5 and 8.0) and,
depending on the strain, at temperatures between 5°C and 42.5°C.
Growth of *C. botulinum* is often associated with an unpleasant
smell which may provide warning of contamination. The pH and
temperature requirements for spore germination are similar to
those for growth. Salt inhibits growth completely if present in
concentrations of 10% or more.

A number of micro-organisms can interfere with the production
of toxin by *C. botulinum*. They include *C. sporogenes*, *Bacterium
linens* and nisin producing strains of *Streptococcus lactis*.
Synergistic effects of lactic acid bacteria and yeast have been
reported, though the former may also exert an inhibitory effect
following their production of peroxides.

4.1.3   The toxins

C. botulinum toxins are heat labile, antigenic proteins. Treatment
with formaldehyde destroys their toxicity without affecting their
antigenicity. The derivatives produced are referred to as toxoids.
Toxoids may be injected safely into suitable animals to produce
antisera for use in therapy.

Most research has been done on the type A toxin. This toxin is
very resistant to acid and fairly resistant to breakdown by pepsin
and trypsin. It is rapidly degraded by alkali and gradually
destroyed by exposure to sunlight and air.

The available evidence suggests that the toxins exert their effects
following absorption from the small intestine into the lymph. The
main effect is paralysis of the efferent autonomic nervous system,
causing death by asphyxia. The site of action is at the synapses of
efferent parasympathetic and somatic motor nerves, i.e. the motor
endplates. Only the cholinergic system is affected. The
fundamental effect appears to be inhibition of acetylcholine
release.

There is some selectivity among the toxins with regard to the
animal species affected. Human poisoning is caused mostly by
types A, B, and E. Type C is involved mainly in poisoning of
birds, small mammals, and horses and type D has been associated
with poisoning of cattle and sheep in South Africa.

4.1.4   Prevention of botulism

To be sure that a given food is safe for consumption at least one of the following criteria must be satisfied. Firstly, the pH of the food should never have been greater than 4.5 after removal from the living plant or animal. Secondly, the salt content of the food should have been at least 10% by weight or the food should have been dried to an equivalent water activity. Thirdly, the food should have been stored at temperatures below 3°C. Fourthly, the food should have been heated to at least 90°c. just before consumption. Fifthly, the food should have been processed in an airtight container at 121°C (250°F) for three minutes. Since these requirements cannot always be fulfilled in practice, the most important step in the prevention of botulism is the avoidance of *C. botulinum* contamination at all stages of food processing.

Where the above criteria cannot be satisfied, e.g. in cooked and processed meats, nitrates may be added to prevent *Clostridium* growth. Since nitrates may react with other substances to form carcinogens in certain circumstances (see Section 4.5.2), there is a very difficult decision to be made in judging how much to use to maximize prevention of botulism while minimizing the risk of causing cancer.

4.2   Staphylococcal Enterotoxins

4.2.1   Staphylococcal food poisoning

Only primates seem to be sensitive to ingested staphylococcal enterotoxins. The symptoms are vomiting and diarrhoea, 2 to 6 hours after eating contaminated food. Other symptoms include salivation, nausea, retching and abdominal cramp. There is no effective treatment to alleviate the symptoms but immunization by injection of the inactive toxoid, formed by formaldehyde treatment of the toxin, may be used as a preventive measure.

Death is rarely a result of staphylococcal poisoning in humans but has occurred in very young children, elderly people with other ailments and following treatment with antibiotics. Antibiotics kill the normal bacterial flora of the intestinal tract, allowing unrestricted growth of the resistant staphylococci, with consequent production of large amounts of enterotoxin. Death is usually the result of shock and the circulatory collapse which may follow severe diarrhoea.

4.2.2   Staphylococcus aureus

Staphylococci are Gram-positive, spherical bacteria which typically occur in grapelike clusters. Their usual habitats are the skin and nasopharynx of animals. *S. aureus* is used as the specific name for pathogenic staphylococci.

Production of the enterotoxins is associated with staphyloccal growth. Production can be inhibited by protein synthesis inhibitors; by inorganic salts, e.g. potassium hydrogen phosphate, potassium chloride, sodium fluoride; by detergents, e.g. Tween 80, sodium deoxycholate; and by agents blocking cell wall synthesis, e.g. penicillin, D-cycloserine, bacitracin. The inhibition by detergents and cell wall synthesis inhibitors suggests that enterotoxin synthesis may be associated with the cell surface since the synthesis of other toxins by staphylococci is not affected.

4.2.3 The enterotoxins

The staphylococcal enterotoxins are antigenic proteins which can withstand heating to 100°C and above without loss of toxicity. As with the *C. botulinum* toxins, treatment with formaldehyde destroys their toxicity without affecting their antigenicity. The toxoids produced in this way may be injected safely into suitable animals to produce antisera. Because of the rapid onset of poisoning these antisera are of no use in therapy, but they have been useful in distinguishing between enterotoxins. Five enterotoxins, designated A, B, C, D and E, have been identified in this way. The type A enterotoxin is the one most commonly responsible for poisoning in humans.

Precisely how staphylococcal enterotoxins exert their effects is not clear. It even remains to be established whether the enterotoxins, or their derivatives, enter the circulatory system or not. Since animals become resistant to enterotoxins after repeated oral doses, and since this is most easily explained on the basis of antibody formation, entry to the bloodstream in a relatively unmodified form seems probable. However, antibodies could not be detected in the bloodstream of monkeys which had been made resistant in this way.

The vomiting response is apparently due to an effect on the abdominal viscera sending a sensory stimulus to the vomiting centre in the brain through the vagus and sympathetic nerves. Enterotoxin diarrhoea has been attributed to inhibition of water absorption from the intestinal lumen or to increased transmucosal fluid flux into the lumen. Possibly both processes are involved.

4.2.4 Prevention of staphylococcal food poisoning

Staphylococcal contamination of food cannot be detected by taste, smell, texture or appearance of the food. Since the enterotoxins can withstand heating to 100°C and above, even normal cooking may not protect the consumer. The only sure means of prevention of staphylococcal poisoning is, therefore, careful food preparation and storage, with every precaution being taken to prevent contamination with *S. aureus*. Since contamination comes mainly from the skin of those involved in food preparation, the key factor here is personal cleanliness.

4.3   Mycotoxins

Illnesses caused by mycotoxins (mycotoxicoses) have often been described in animals, but relatively infrequently in human beings. Such illnesses are always associated with a specific food which shows signs of fungal growth. They are often seasonal and are nontransmissible and unresponsive to drug and antibiotic treatment.

4.3.1   Ergot

Ergot is a mixture of alkaloids produced by species of the genus *Claviceps*, which grows on rye and wild grasses. The most studied species of this genus is *Claviceps purpurea*. Ergot causes the mycotoxicosis known as ergotism, which was once common in human populations but has now virtually disappeared because of improved grain harvesting and storage.

The ergot alkaloids may be divided into two groups—the acid amide derivatives of lysergic acid and its stereoisomer, isolysergic acid, and the clavines. These alkaloids can exert a wide range of effects. Peripherally, they can cause vasoconstriction and contraction of the uterus. Acting on nerves, they can antagonize the effects of epinephrine (adrenaline) and serotonin (5-hydroxytryptamine). In the central nervous system, they can reduce the activity of the vasomotor centre while stimulating the sympathetic regions of the midbrain, especially the hypothalamus.

Though ergotism has been almost eliminated, the ergot alkaloids are still important as starting material for the production of drugs, many of which are useful in treating human illness. Amongst these drugs is lysergic acid diethylamide (LSD), one of the most powerful hallucinogenic agents known.

4.3.2   Aflatoxins

The aflatoxins are probably the most dangerous mycotoxins. Aflatoxins have been shown to be responsible for hepatomas (liver tumours) in hatchery raised trout and for the deaths of turkeys, ducklings, chicks, and young pheasants. Four principal aflatoxins have been identified and have been designated as B_1, B_2, G_1 and

G_2. These toxins are produced by various fungi, including *Aspergillus flavus, Aspergillus parasiticus, Aspergillus ostianus* and other species of *Aspergillus* and *Penicillium*.

Aflatoxins have been found in mouldy peanuts, cottonseed, coconuts and corn. They have even been found in dried fermented fish and have been identified in the milk of animals eating aflatoxin contaminated feed. Almost all reports of their occurrence refer to stored foodstuffs. Their production in the field seems to be rare.

The most characteristic effect of aflatoxins is liver damage, with proliferation of bile duct cells in cases of acute lethal poisoning, or with tumours following chronic exposure. Aflatoxins may also affect insect reproduction adversely and inhibit seedling germination and the growth of micro-organisms.

Prevention of aflatoxin synthesis is the best way of eliminating its effects. mechanical damage of crops during harvesting should bc minimized and care should be taken that the moisture content of susceptible crops following harvesting is kept below the level necessary for fungal spore germination. Where fungal contamination has occurred, contaminated crops can sometimes be separated out using photoelectric sorters. This technique has been used for peanuts, since contaminated peanuts are usually darker in colour than clean nuts. There is as yet no satisfactory process for detoxification of contaminated foods.

4.3.3   Other mycotoxins

Apart from the aflatoxins, there are a number of less thoroughly researched mytotoxins. *Fusarium sporotrichioides, Fusarium poae, Cladosporium epiphyllum* and *Cladosporium fagi* produce sporofusariogenin, poaefusariogenin, epicladosporic acid and fagicladosporic acid respectively. These may cause an illness called alimentary toxic aleukia which has been reported by Russian workers. Its symptoms include leukaemia, agranulocytosis, necrotic angina, a tendency to bleed easily, sepsis and aregenerative exhaustion of the bone marrow. It is associated with eating fungal contaminated grain, e.g. wheat, rye, oat, buckwheat and millet, which has been overwintered in the field.

Another toxin is produced by *Aspergillus ochraceous* which occurs in soil, moist cereals, legumes, spices and katsuobushi (Japanese fermented fish). The toxin is called ochratoxin and occurs in three forms, A, B and C. Ochratoxin A has the same degree of acute toxicity as aflatoxin B_1. Ochratoxins B and C are nontoxic at a thousandfold higher dosage in tests of acute lethality. These ochratoxins are thought to be carcinogenic.

Sporidesmin is the toxin produced by the saprophytic mould,

Pithomyces chartarum, growing on grasses. It causes obstructive damage to bile ducts and facial eczema in sheep and cattle. Zearalenone (F-2 oestrogenic factor) is a toxic substance produced by *Fusarium zea* growing on corn. The effects of the compound include vulvar hypertrophy, preputial enlargement in castrated males and prominent mammary glands in both sexes of animals fed contaminated grain.

Sclerotinia sclerotiorum, growing on celery, is responsible for the production of 4, 5′, 8-trimethylpsoralen and 8-methoxypsoralen. However, it is not clear whether these compounds are produced by the fungus or by the celery in response to the fungal infection. The characteristic effect of these toxins is a blistering lesion of skin after it has come in contact with contaminated parts of the plant and after subsequent exposure to sunlight, which is necessary for the lesion to appear.

Red clover hay containing *Rhizoctonia leguminicola* has been shown to cause cattle and sheep which eat it to salivate excessively and develop diarrhoea, bloating and stiff joints. Sometimes this culminates in death. The mycotoxin causing these symptoms has been isolated and named slaframine. It is also toxic to pigs, chickens, guinea pigs, rats and mice.

Various penicillia have been implicated in the production of the so-called yellow rice toxins. These include *Penicillium islandicum, P. citrinum, P. toxicarum, P. citreoviride,* and *P. rugulosum. P. islandicum* produces islanditoxin and luteoskyrin. Islanditoxin causes rapid death with severe liver damage and haemorrhage. Luteoskyrin at high doses causes centrolobular necrosis of the liver in mice, accompanied by fatty metamorphosis of the other hepatocytes. It is thought tha it may cause liver tumours at lower dosages. *P. toxicarium, P. ochrosalmoneum* and *P. citreoviride* produce citreoviridin, a yellow fluorescent compound which causes paralysis and respiratory failure. It is inactivated by ultraviolet light and, therefore, contaminated rice becomes detoxified when exposed to the sun for two days. *P. citrinum* produces citrinin, as do certain other penicillia, *Aspergillus terreus* and *A. candidus.* Citrinin and citreomycetin, a related compound produced by *Citromyces* and some penicillia, cause kidney damage following chronic exposure.

Stachobotrys alternans growing on hay produces a toxin which affects horses, sheep, calves, pigs, dogs, rabbits, guinea pigs, chicks and mice. The toxin causes severe haemorrhage and other damage to many tissues, which often leads to death. Superficial contact causes severe dermatitis in human beings and may cause serious pharyngitis. *Penicillium rubrum* growing on corn produces rubratoxin. This causes haemorrhaging and liver and kidney damage in cattle, pigs, dogs, rabbits, guinea pigs, mice and chickens.

The mycotoxins described above represent only a selection of those that have received most attention. To a large extent, the dangers posed by these toxins can be minimized by proper handling and storage of food, i.e. by preventive measures. Much more, however, needs to be known about the diagnosis of mycotoxicoses and about the means of alleviating their ill effects, since it is likely that even the best preventive measures will not succeed in eliminating them completely.

4.4 Algal Toxins

The algal toxins that have attracted the most attention are associated with three phyla: the Pyrrophyceae, the Cyanophyceae and the Chrysophyceae. Of these, the Pyrrophyceae are predominantly marine organisms, often classified as animals, Protozoa, Mastigophora; the Cyanophyceae or blue-green algae are freshwater organisms, sometimes classified as bacteria, Cyanobacteria; and the Chrysophyceae favour brackish water. Only the Pyrrophyceae have posed a serious poisoning threat to man. The Cyanophyceae have caused the deaths of fish, cattle and water birds, and the Chrysophyceae have caused serious fish mortality.

4.4.1 Pyrrophyceae

The most impressive manifestations of the dinoflagellates belonging to the phylum Pyrrophyceae are the 'red tides' that occur from time to time, mostly in waters north or south of 30° latitudes. The dinoflagellates may reach concentrations of up to 50 000 cells per cm^3, and may be clearly visible at night because of their fluorescence. These red tides affect man because the toxins they carry can be accumulated safely by mussels and clams, which are subsequently eaten by human beings. The consequences for the human consumer are rapid. Numbness spreads from the lips, tongue and finger tips, and death from respiratory paralysis may follow in 2 to 12 h. Salt and alcohol reduce the effectiveness of the poison but there is no antidote. However, survival for 24 h is normally followed by complete recovery.

Four species of dinoflagellate, *Gonyaulax catenella, G. tamarensis, G. acatanella* and *Pyridinium phoneus*, have been shown to produce toxins which can be lethal to *Homo sapiens*. Other species produce toxins as well but these have not been implicated in fatal human poisoning. *G. monilata* produces a toxin which is fatal to fish and inhibits oyster feeding; *Gymnodinium breve* produces a toxin which is fatal to fish, chicks and mice; *Gym. veneficium* toxin is fatal to fish and mice; *Exuviaella mariae-lebouriae* toxin causes liver and kidney damage in animals. In general, these toxins cause vomiting, diarrhoea, muscular weakness and dizziness in human beings.

During a red tide, mussels accumulate the toxin in the hepatopancreas but, once the red tide is over, the toxin disappears from the mussels within two weeks. In clams, accumulation occurs in the siphon and here the toxin is persistent. It may not disappear from the clam siphon for up to a year after accumulation ceases.

The toxin from *G. catenella* has been named saxitoxin. It has been purified and shown to be a derivative of tetrahydropurine. Its fundamental effect is to block the propagation of impulses in nerves and skeletal muscle.

4.4.2 *Cyanophyceae*

Only species from three Cyanophyceae genera, i.e. *Microcystis*, *Anabaena* and *Aphanizomenon*, have been studied in any detail with regard to toxin production. The toxins appear to be predominantly endotoxins. *Microcystis aeruginosa* produces a toxin called Microcystis FDF (fast death factor) which is a small cyclic peptide. *Anabaena flos-aquae* produces a toxin which seems to be a low molecular weight tertiary amine or alkaloid. The toxin of *Aphanizomenon flos-aquae* has been studied but not characterized.

Blooms of Cyanophyceae are associated with excessive eutrophication of lakes and ponds caused either by leaching of fertilizers from agricultural land or by input from domestic and other sewage (see Section 9.2.3). As these blooms die, the toxins are released and animals may be poisoned by drinking the contaminated water. Although these toxins have not caused any human fatalities, they have been blamed for causing dermatitis, gastroenteritis and respiratory disorders in human beings. The toxicity of contaminated water shows diurnal fluctuations. When photosynthesis is maximal, around mid-day, dissolved oxygen levels are high and the pH of the water may reach 9.5. This lowers the toxicity of the water because all the Cyanophyceae toxins are alkali labile, especially in the presence of high oxygen concentrations and in warm water (20°C and above). At night, only respiration occurs, the oxygen level of the water drops and the pH falls to about 6.5. The stability of the toxin under these conditions makes for maximal concentrations. Thus, any steps to destroy Cyanophyceae blooms should be taken in the middle of the day when the toxin released will break down quickly and so pose a minimal threat to zooplankton and fish. Decomposition of the Cyanophyceae not only releases toxins but also leads to a depletion of oxygen, accompanied by production of hydroxylamine and hydrogen sulphide which are themselves highly toxic.

4.4.3 Chrysophyceae

The Chrysophyceae are yellow or brown single cell flagellate algae. One member of the phylum, *Prymnesium parvum*, has been thoroughly studied because it excretes a toxin which has been responsible for the periodic mass deaths of fish in brackish waters. This toxin has been isolated and appears to be a proteolipid. It has haemolytic, cytotoxic and bacteriolytic activity, in addition to its ichthyotoxic effect. The ichthyotoxic effect seems to be associated with the lipid moiety, since it is not destroyed by the action of the proteolytic enzyme papain, which eliminates haemolytic activity. Ichthyotoxicity requires the presence of cations at pHs above 7. Calcium, magnesium, polyamines, cationic detergents and even antibiotics such as streptomycin and neomycin, have been shown to enhance toxicity. The cations appear to compete for sites on the toxin. A high concentration of a cation of low activity may displace another of high activity, thus lowering the toxicity. Visible and ultraviolet light can inactivate the toxin. The toxin appears to act by causing a reversible increase in gill permeability which permits it to enter the bloodstream of the fish and act on the nervous system.

Prymnesium parvum is never found in water containing less than 0.12% sodium chloride. In the laboratory, it will grow in water with salt concentrations up to three times that of seawater. Control of the organism in fish ponds depends upon detection of the toxin at very low concentrations, followed by elimination of the alga by addition of aqueous ammonia which causes the cells to swell and lyse. At the concentrations used, the toxicity of ammonia to other forms of life is negligible. Detection of the toxin involves a bio-assay using minnows in the presence of 3,3'-diamino-dipropylamine as cationic activator at pH 9. This assay can detect *Prymnesium* toxin at levels equal to 4% of the concentration lethal to carp.

Ochromonas is another member of the Chrysophyceae which produces an ichthyotoxin. In addition, *Phaeocystis pouchetii* in the mucilaginous colonial form is toxic to herring fry. Other *Phaeocystis* spp. produce acrylic acid which is accumulated by euphausiids which are eaten by penguins. The acrylic acid may then exert an adverse effect on the gut microflora of the penguins.

4.5   Other Toxic Products of Microbial Metabolism

Almost every substance which enters the environment will be metabolized by one or more of the micro-organisms which encounter it. The products of this metabolism are often simple essentially nontoxic compounds, e.g. carbon dioxide and water, but this is by no means always the case. For example, the herbicide propanil which has low toxicity to animals may be converted by soil micro-organisms to the mutagens 3,4-dichloroaniline and 3,3',4,4'-tetrachloroazobenzene which, in turn, may cause mutations of soil pathogens to more virulent strains. More often, the products of concern are simple compounds which are harmful only when produced locally in high concentrations. These are the compounds which will be discussed in this section.

4.5.1   *Ammonia*

Ammonia is formed during the microbial decomposition of plant and animal remains under alkaline conditions. It may also be formed by the breakdown of urea applied as a fertilizer. Ammonia produced in this way can reach concentrations which are sufficient to harm plant root systems and soil-dwelling organisms.

4.5.2   *Nitrate, nitrite and amines*

Nitrate is another product of the microbial breakdown of organic matter. Under aerobic conditions, at pHs near neutral, ammonium ions are oxidized to nitrate ions. Although nitrate is fairly harmless to animals, its derivatives formed by gut micro-organisms may be dangerous. Firstly, there is nitrite. Nitrite reacts with haemoglobin to convert it to methaemoglobin, by oxidizing the ferrous iron in haemoglobin to the ferric state. Methaemoglobin is incapable of carrying oxygen and so its formation is associated with oxygen deficiency in the tissues. Treatment with ascorbic acid protects haemoglobin and permits total recovery if the effects have not been too severe. Infants are particularly susceptible to nitrite because oxidation of foetal haemoglobin occurs very readily and because the protective enzyme, methaemoglobin reductase, develops only gradually after birth. Because of this danger, the World Health Organization has recommended that water for human consumption should never have more than 10ppm nitrate. Any nitrite present is in equilibrium with nitrate and at a much lower concentration.

Other compounds that may be formed from nitrate by gut microflora are carcinogenic nitrosamines. In oxygen-deficient natural waters containing nitrate and ammonia, hydroxylamine may be produced by microbial action. Hydroxylamine is a potent mutagen. During silage production from nitrate-rich plants, nitric oxide may be formed and converted to nitrogen dioxide by atmospheric oxidation. Cases of farmers being poisoned as a result of this have been reported. (For further information on the toxicity of nitrogen oxides see Section 7.5.)

Microbial decomposition of protein in food can lead to the production of toxic amines, referred to collectively as ptomaine. This used to be a common cause of food poisoning, and may still be so where elementary hygiene is not practised and where food preservation is poor.

4.5.3 Carbon monoxide and carbon dioxide

The carbon in organic matter is mostly converted to carbon monoxide and carbon dioxide by micro-organisms. The risks associated with carbon monoxide are discussed in Section 7.1 High carbon dioxide levels can cause injury to roots in poorly drained or compacted soils. However, in enclosed waters it may contribute to excessive eutrophication (see Section 9.2.3).

4.5.4 Sulphur derivatives

Sulphur in organic matter and in inorganic sulphate may be converted by micro-organisms into a range of compounds, most commonly hydrogen sulphide. Hydrogen sulphide production is most marked in lake and ocean bottom sediments, swamps, marshes, estuarine waters, raw sewage and industrial effluents. Whether free hydrogen sulphide is released depends upon the availability of cations which can react to form sulphides. Higher plants are very sensitive to hydrogen sulphide. It has been responsible for considerable damage to rice crops and to fruit trees. At sublethal levels it may make roots susceptible to parasitism by the pathogen, *Phytophthora*. In water very low concentrations (<0.1 ppm) are toxic to fish eggs and fry. This may be a frequent cause of mortality at the mud-water interface. Other potentially dangerous products of microbial sulphur metabolism are sulphuric acid, sulphur dioxide, carbonyl sulphide, dimethyl sulphide, methane thiol and ethane thiol. Sulphuric acid is not only directly toxic because of its acidity but may solubilize metals such as zinc, aluminium, copper, iron, nickel, arsenic and cadmium till they reach toxic levels. This is a danger particularly associated with drainage of mines in which the sulphide ores provide substrates for sulphur oxidizing autotrophs (see Section 6.14).

5 Pesticides and Herbicides

Pesticides are substances used to control organisms which may adversely affect public health, or organisms which attack food and other materials essential to mankind. Such organisms include vermin, insects, nematodes and fungi. Herbicides are substances used to eliminate unwanted plants in agriculture, horticulture and in the maintenance of road verges and railway tracks. Herbicides have also been used in jungle warfare in Vietnam to destroy tropical forests.

Pesticides and herbicides have made possible great increases in food production and improvements in human health. Where harm has occurred, this has largely been due to ignorance of the properties of these chemicals and to misuse, i.e. wrong quantity, application or timing. Since it is impossible to know every property of every pesticide or herbicide, it is inevitable that some environmental damage will occasionally result from their use. However, this is no argument against their use, so long as the benefits can be clearly shown to outweigh the damage. The important thing is to be aware that there is a finite risk of damage, to minimize it by appropriate toxicity testing of chemicals before they are applied on a large scale and to ensure that the application is carried out correctly. When pesticides are applied on a large scale, a constant watch must be kept for environmental damage so that corrective steps may be taken as soon as possible.

In this chapter, some of the hazards associated with widely used pesticides and herbicides will be considered. One hazard which will not be covered is mutagenicity since interpretation of the currently available data is very difficult. For information on the mutagenicity of pesticides, herbicides and many other compounds, the reader is referred to the review of Malling and Wassom (1977).

In general usage pesticides are referred to by their trivial names and this practice will be followed throughout this chapter. However, the full chemical names of all pesticides discussed can be found in Appendix 4 at the end of the book.

5.1 Rodenticides

Anticoagulant rodenticides such as diphacin, fumarin, PMP, pival, prolin and warfarin are widely used. Rodents which eat small amounts of these chemicals over several successive days eventually die from internal bleeding. The hazard from these anticoagulants to human beings and pets is regarded as minimal. Unfortunately, some rodents can develop a tolerance to these substances and this may necessitate the use of more toxic compounds which cause death following ingestion of a single dose. These compounds are not usually available to the general public because they are lethal to many animals other than rodents. Hence they must be used only with special precautions. Such compounds include antu, fluoroacetamide, red squill, sodium fluoroacetate (Compound 1080), strychnine and derivatives and thallium sulphate. In addition to this group there is norbormide which is acutely toxic to members of the genus, *Rattus*, but has little effect on other species, including *Homo sapiens*.

Gaseous rodenticides are sometimes used to kill rodents in their burrows or to fumigate buildings. The gases used include hydrogen cyanide, carbon monoxide, carbon disulphide, chloropicrin, sulphur dioxide and methyl bromide. All these gases are toxicants to many organisms and proper precautions must be taken to protect those applying the gases and nontarget organisms.

5.2 Insecticides

5.2.1 Chlorinated hydrocarbons

Chlorinated hydrocarbons (Fig. 5.1) are the most prevalent pesticides in the environment because of their wide use and persistence. The most persistent of these compounds are DDT and derivatives such as DDD and DDE. DDT has an environmental half life of ten years or more, while DDE can persist for decades. Almost as persistent are lindane (BHC) and heptachlor. Less persistent are aldrin and dieldrin, but even these require 2½ years in the soil for 95% degradation. All of the compounds mentioned are currently in use throughout the world, though DDT, aldrin and dieldrin have been banned for most purposes in the United States, and their use has been discouraged in the United Kingdom and a number of other countries. Reaction against the use of these compounds has followed awareness of their toxicity to nontarget organisms and the discovery that some target insect populations were becoming resistant. However, substantial quantities of chlorinated hydrocarbons are still used in wood preservation and in sheep dips where there is, at present, no satisfactory substitute. The most extensive use of DDT now occurs in tropical countries because it is cheap, persistent, generally effective and demonstrable harm to human beings has been minimal. There can be no

Structure	Trivial name	Scientific name
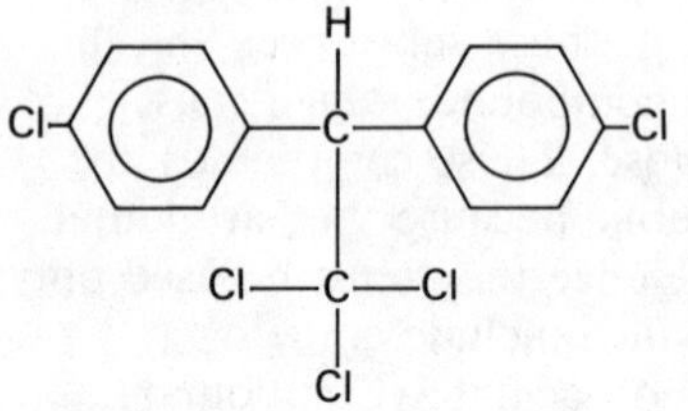	DDT	1,1,1-Trichloro-2,2-bis (*p*-chlorophenyl) ethane
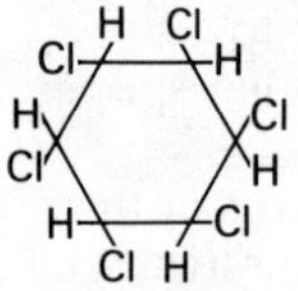	Lindane (BHC)	1,2,3,4,5,6-hexachloro-cyclohexane (γ-isomer)
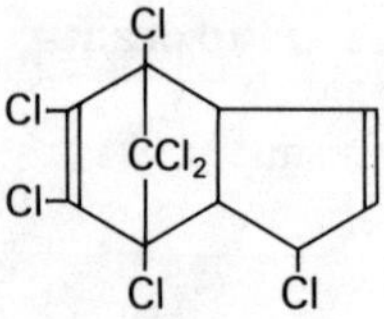	Heptachlor	1,4,5,6,7,8,8-heptachloro-3α,4,7,7α-tetrahydro-4,7-endomethanoindene
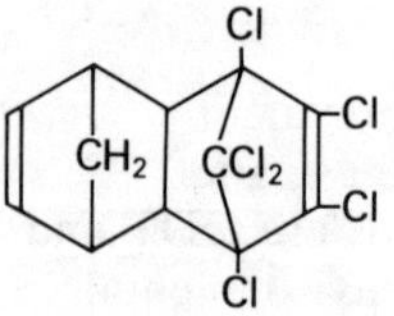	Aldrin	1,2,3,4,10,10-hexachloro-1,4,4α,5,8,8α-hexahydro-1,4-*endo-exo*-5,8-dimethanonaphthalene.
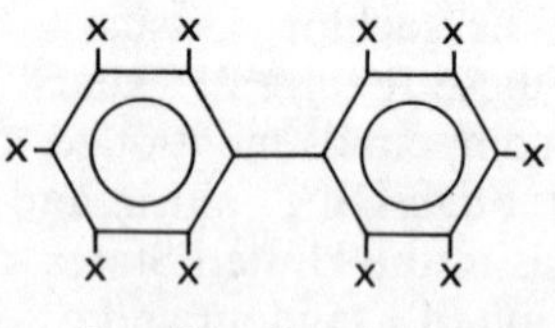	PCB	Polychlorinated biphenyl

x = Possible site of chlorine atom

Fig. 5.1 Structures of selected chlorinated hydrocarbons.

doubt that in these countries DDT has alleviated human suffering by reducing insect-borne disease and destruction of crops. Until a suitable alternative is available, its use will continue. At present, possible substitutes are much more expensive, less persistent and more toxic to human beings.

Most of the problems caused by effects of DDT on nontarget organisms are the result of its being used in excessive amounts. The excess finds its way into ponds, lakes, rivers and, ultimately, the sea. In some cases, DDT is applied directly to fresh water insect habitats; in others, accidental spraying of such habitats may occur. Otherwise, DDT may enter waterways in surface run-off, or be washed out of the atmosphere in rain or snow. Despite this, the concentration in natural waters is low. However, many plants and animals tend to accumulate DDT and its more persistent derivative DDE (Fig. 5.2). For example, oysters can concentrate DDT from 1 ppb in seawater to 700 ppm in their bodies. Water fleas can concentrate DDT from 0.5 ppb in the surrounding water to 50 ppm in their bodies. This process of accumulation can continue through food chains until harmful levels are reached in the ultimate predators (see Section 9.2.1). The effect is, of course, greater in the case of DDE owing to its greater persistence and prevalence. Accumulation occurs largely because of the affinity of DDT and DDE for fats. This leads to their localization in animals in adipose tissue where the turnover rate is low. Stress may cause mobilization of adipose tissue. The subsequent release of the accumulated residues may lead to toxic levels in the blood of animals even after exposure to the pesticides has ceased.

The harmful effects of chlorinated hydrocarbons were first noted in birds of prey where reproductive failure led to a marked decline

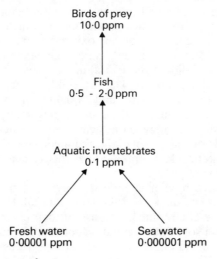

Fig. 5.2 A simplified diagram showing food chain accumulation of DDT. (Concentrations are taken from Edwards, C. A., 1973, *Persistent Pesticides in the Environment*, 2nd edn, CRC Press).

in many populations. This was partly due to impaired calcium metabolism which resulted in fragile eggs with abnormally thin shells, and partly due to behavioural changes which favoured egg breakage. This has proved to be a problem in battery chicken farming too, where sawdust from wood treated with chlorinated hydrocarbons has been used as litter. Harmful effects on fish have also been noted, including behavioural changes leading to reproductive difficulties, increased mortality among the young and, in some cases, acute toxicity to adults. Further, small concentrations of DDT (0.01 ppm) have been shown to reduce photosynthesis in marine plankton, while even smaller concentrations (1 ppb) in seawater can kill many brine shrimps (*Artemia salina*) within a few weeks.

Another insecticide in this group which has caused environmental damage in the USA is kepone. This compound was used domestically as an ant or cockroach poison, but its production was stopped when it was shown to have caused brain and liver damage, sterility, slurred speech, loss of memory and eye-twitching in workers who had been exposed to it. Matters were made worse by leakage of kepone from the production plant into the main sewer system of Hopewell, Virginia and, therefore, into the James River. As a result, fish and shellfish stocks were contaminated and the Governor of Virginia closed more than 100 miles of the river to commercial fishing until such time as residue concentrations should fall to a safe level. Losses to fishing and related industries were estimated at about $4 million for the period July 1975 to December 1976. Another insecticide, mirex, which is chemically identical to kepone except for one oxygen atom, is used to control ants, but there is concern about its possible carcinogenicity.

Polychlorinated biphenyls (PCBs) are structurally similar to DDT and have similar effects. They are stable at temperatures up to 800°C, resistant to acids, bases and oxidants, and only sparingly soluble in water. Because of these properties, many uses have been found for them. They may act as heat transfer fluids, insulators, hydraulic fluids or plasticizers, and are incorporated in paints, adhesives, gasket sealers, brake linings, fluorescent lamps and carbonless copying paper.

Polychlorinated biphenyls enter the environment accidentally or through poor waste disposal. Like DDT and DDE, they are persistent in the environment because of their chemical stability. Also, their lipid solubility facilitates food chain amplification (see Section 9.2.1). Their acute effects are not a serious problem but their chronic effects are so similar to those of DDT and DDE that it is likely that they act synergistically. Because of this, their use is now carefully controlled in most countries and largely restricted to

closed system applications. However, care must still be taken to
avoid accidents. This was clearly demonstrated in 1973 when
225–450 kg (500–1000 lb) of the related polybrominated biphenyls
(PBBs) were accidentally added instead of magnesium oxide to
livestock feed in Michigan, USA. This resulted in contamination of
farm animals making necessary the destruction of approximately
29 800 cattle, 5920 pigs, 1470 sheep and 1.5 million chickens. Also
rendered unsaleable were more than 880 tonnes (865 tons) of feed,
8.1 tonnes (8 tons) of cheese, 1.2 tonnes (1.2 tons) of butter, 15.4
tonnes (15.2 tons) of dry milk products and nearly 5 million eggs.
It is impossible to know how many people have suffered sublethal
poisoning as a result of eating contaminated meat, eggs and milk,
but it may be a very considerable proportion of the population of
Michigan. Amongst the possible effects of poisoning with PBBs
have been identified teratogenicity and carcinogenicity in rodents,
and porphyria and reduced fertility in birds. The origins of this
disaster seem to have been due to a shortage of bags prelabelled in
red so that the PBBs, marketed as the fire retardant 'Firemaster',
were for a time packed in plain brown bags on which the trade
names were stencilled in black. Magnesium oxide, a normal feed
additive, marketed as 'Nutrimaster' and similar in appearance to
Firemaster, was produced in the same plant and packed in
identical brown bags, differing only in the stencilled name.
Somehow, ten to twenty 50 lb bags of Firemaster were included in
a truck-load of Nutrimaster. Because of the superficial similarity of
contents and packaging this was not noticed. As a result, animal
and poultry feed throughout Michigan was contaminated. To make
matters worse, some farmers sold contaminated animals to
rendering plants to be processed into livestock feed, dog and cat
food as well as tallow and grease used in soaps, shampoos etc. This
accident illustrates the importance of clear labelling of toxic
substances and of constant vigilance to spot poisoning before it
becomes widespread.

5.2.2 *Organophosphates*

The organophosphate insecticides include malathion, diazinon and
parathion. They are readily metabolized by mammalian enzymes
and rapidly degraded in the environment. Thus, any environmental
damage caused by these compounds will tend to be localized in the
area of application. However, these compounds are related to the
nerve gases developed for use in war and can be lethal to many
different organisms, including humans. Hence, anyone applying
these compounds should wear protective clothing, including
goggles, rubber gloves and even a respirator. One should avoid

treating areas which are important habitats for wildlife.

The principal action of the organophosphate compounds is to inhibit acetylcholine esterase. Thus, the action of acetyl choline released at the nerve synapses ceases to be finite with each nerve impulse. Amongst the consequences are tremors in involuntary muscles, convulsions and death.

5.2.3 Carbamates

Carbamate insecticides, also used as molluscicides, fungicides and herbicides, include carbaryl (sevin), baygon, temik and zectran. They are even less persistent than the organophosphates and less harmful to man. As always, however, they may cause local environmental problems if used carelessly. Carbaryl, for example, is very toxic to bees. Like the organophosphates, the carbamates act by inhibiting acetyl choline esterase.

5.2.4 Pyrethroids

Pyrethroids are the insecticidal constituents of pyrethrum flowers (*Chrysanthemum cineriaefolium* and *Chrysanthemum coccineum*). Five insecticidal compounds, pyrethrins I and II, cinerins I and II, and jasmolin II, are present in the achenes of the flowers from which they may be extracted with organic solvents. These naturally occurring pyrethroids are unstable to the action of sunlight, air, moisture and alkalis, and are rapidly degraded after application.

Pyrethroids have a low toxicity to mammals because they are rapidly metabolized to harmless substances. However, they can cause severe allergic dermatitis and systemic allergic reactions. Large amounts may cause nausea, vomiting, headache and other disturbances of the central nervous system.

The high cost of extraction of natural pyrethroids has led to the production of similar synthetic compounds, of which the best known are the allethrins. The synthetic compounds have a wide range of properties and it is impossible to generalize about them.

5.3 Nematicides

5.3.1 Against plant parasitic nematodes

The nematicides can affect large areas of land. The main nematicides used are 1,3-dichloropropene (telone) and 1,2-dibromoethane, both of which are liquids applied directly to the soil. Because they are phytotoxic they can only be used in the absence of growing crops. These compounds are also very toxic to mammals. They irritate the skin, eyes and mucous membranes, damage the liver, and ultimately cause death from severe lung injury. Other soil nematicides in use are sodium N-methyldithiocarbamate in aqueous solution (see Section 5.2.3.), O-2, 4-dichlorphenyl-O-diethylphosphorothioate (V-C 13 nemacide) as an emulsion (see Section 5.2.2), calcium cyanamide, urea, and 3,5-dimethyl-tetrahydro-1, 3, 5, 2H-thiadiazine-2-thione (mylone). The last three chemicals are solids and must be thoroughly mixed with the soil to be effective. All of these compounds are toxic to mammals and probably to most other animals as well.

Fumigants, such as methyl bromide, may be used in confined spaces, or under polythene sheeting, where only small areas require to be treated. Methyl bromide is very toxic and exposure of mammals to 50 mg per body weight can cause death.

Seeds and bulbs may sometimes be treated directly with dilute formaldehyde solution in water (0.5 to 1.0% weight to volume), with or without a trace of detergent. It is unlikely that this constitutes an environmental hazard.

5.3.2 *Against mammalian parasitic nematodes*

Whether the drugs used in the treatment of nematode infection in human beings and domesticated animals constitute an environmental hazard does not seem to have been assessed. However, since at least 200 million people throughout the world are thought to be infected with nematodes, and since presumably a similar proportion of the domesticated animal population is at risk, the use of nematicidal drugs must be considerable. In consequence, appreciable amounts of these compounds or their derivatives must enter the environment. It is, therefore, justifiable to describe briefly here the principal drugs involved.

Diethylcarbamazine is the most effective drug against filarial nematodes. It is not highly toxic but it can cause nausea, vomiting and headache in humans, and muscle tremors and convulsions in dogs. Thiabendazole is the most widely used drug for the control of gastro-intestinal nematodes, other than *Trichuris* spp., in ruminants. It is also used on citrus fruit as a fungicide. Its acute toxicity to mammals is low. Phenothiazine is often used against intestinal nematodes in sheep and chickens. It is not used on cattle, horses or humans, because of its many toxic effects on these species. Piperazine is the drug of choice for the treatment of oxyuriasis and ascariasis in both humans and other mammals, since it is highly active against the nematodes responsible but almost nontoxic to mammals at the therapeutic dose levels.

5.4   Fungicides

Probably the most widely used agricultural fungicides are Bordeaux mixture and sulphur. Bordeaux mixture is made by combining a solution of copper sulphate with a suspension of calcium hydroxide, usually in equal amounts by weight. Bordeaux mixture may have harmful side effects on treated plants, e.g. it may cause accelerated transpiration and, hence, skin russeting of apples, and premature defoliation of peach trees. Other effects may be associated with localized concentrations of copper (see Section 6.5). Sulphur has low toxicity but it may cause irritation to the skin, eyes and respiratory tract. It may also be converted to hydrogen sulphide (see Section 4.5.4) which is a very powerful poison.

Organic fungicides are replacing Bordeaux mixture and sulphur in many applications. Organomercurial compounds (see Section 6.2) are used as seed dressings. Dithiocarbamates (see Section 5.2.3) such as ferbam, ziram, nabam, zineb and maneb, are widely applied to vegetables and ornamental plants. Other organic fungicides in use are thiram, captan, glyodin, chloranil, dichlone, dodine (cyprex) and karathane. Of these, thiram, dodine and karathane are appreciably toxic to humans. Captan, glyodin, chloranil and dichlone have low toxicity for humans because they are poorly absorbed from the gut.

Hexachlorobenzene, which has been used as a fungicidal seed dressing, has a relatively low acute toxicity for mammals but can cause serious delayed effects. This was seen most dramatically in Turkey in 1955 when up to 5000 people were poisoned by eating hexachlorobenzene-treated grain. Their symptoms included enlarged livers, abnormal light sensitivity, weight loss and abnormal hair growth, particularly on the face. This last symptom gave the epidemic its name of monkey face disease. It is worth noting that hexachlorobenzene may be an impurity in other pesticides, at levels up to 10% of the total. It is chemically stable and, therefore, persistent. Measurable amounts are released into the environment during the manufacture of perchlorethylene.

The antibiotic, cycloheximide, has been used to treat cherry leaf spot. This antibiotic is extremely toxic, causing gastro-intestinal lesions, coma and death. Sympathomimetic drugs may alleviate the symptoms and hydrocortisone or adrenal cortical extract may prevent death.

As well as the fungicides which are applied directly to plants or seeds, there are others used to clear fungi from the soil before planting crops. Many of these, e.g. formaldehyde, chloropicrin and methyl isothiocyanate, are phytotoxic but their volatility enables

them to escape from the soil before planting commences.
However, these compounds are also highly toxic to animals and
this should be borne in mind when their application is being
planned. Pentachloronitrobenzene is a soil fungicide which appears
to have low phytotoxicity and it has been applied directly to soil in
which cruciferous crops and lettuce are growing. However, it is
moderately toxic to animals and accumulates in body fat. Hence,
there is a potential long-term risk similar to that from chlorinated
hydrocarbons (see Section 5.2.1).

5.5 Herbicides

The toxicity of herbicides to plants has already been discussed in
Section 3.5. In many cases their toxicity to animals is low, but it is
rarely negligible. Repeated doses of diuron have been shown to
produce anaemia and methaemoglobinaemia in rats. A number of
people have died following accidental ingestion of paraquat.
Others have died following excessive exposure to derivatives of
2,4-dinitrophenol. Even the widely used chlorophenoxy acid
herbicides can cause eye irritation and gastro-intestinal
disturbances in animals exposed to large amounts. One of the
chlorophenoxy acid herbicides, 2, 4, 5-T may contain appreciable
quantities of dioxin as an impurity. Dioxin is the name given to the
family of chemicals related to 2,4,7,8-tetrachlorodibenzo-p-dioxin
(TCDD). This family has been shown to be strongly teratogenic
though its most obvious effect is a skin disease called chloracne. It
is also environmentally stable and may accumulate. The United
States Environmental Protection Agency has now banned most
crop related uses of 2,4,5-T, its use on cattle pasture has been
questioned and animals must not be grazed on treated areas within
two weeks of slaughter.

5.6 Inorganic Pesticides and Herbicides

In the past, various compounds of zinc, copper, arsenic, mercury
and other metals have been used as pesticides. Sulphuric acid,
sodium arsenite, sodium chlorate, sodium penta- and metaborate
and sodium thiocyanate have been used as herbicides. Some of the
herbicides, e.g. sodium chlorate, are still used for total ground
clearance but, in general, the use of inorganic pesticides and
herbicides is much less than it was. This is a welcome development
because all these inorganic compounds are highly toxic to a wide
range of organisms and the metallic derivatives may leave
persistent toxic residues in the soil (see Chapter 6). Nevertheless,
it may still be necessary to use such compounds where pests have
developed resistance to the more selective organic pesticides.

6 Toxic Metals

Trace metals in the geological sense are the main source of metal toxicity problems in the environment since most organisms are not adapted to deal with them when they occur locally at high concentrations. In geological terms, trace elements are defined as those occurring at 1000 ppm or less in the earth's crust. Only twelve elements do not come within this classification, i.e. oxygen, silicon, aluminium, iron, calcium, sodium, potassium, magnesium, titanium, hydrogen, phosphorus and manganese. The trace metals may be divided into those that are 'heavy', with densities greater than 5 g cm^{-3}, and those that are 'light', with densities less than 5 g cm^{-3}. Many trace metals are essential at low concentrations for normal life.

Excessive levels of trace metals may occur naturally as a result of normal geological phenomena such as ore formation. Weathering of rocks, leaching or, in the case of mercury, degassing may make these metals available to the biosphere. Humans release more of the metals by burning fossil fuels, mining, smelting, discharging industrial, agricultural and domestic waste, and by deliberate environmental application in pesticides. Once made available to the environment, metals are not usually removed rapidly, nor are they readily detoxified by metabolic activity. As a result they accumulate. Thus, their release into the environment must be carefully monitored and controlled.

Defining the main problem metals is not easy. The United States Environmental Protection Agency has defined beryllium (a light trace metal) and mercury (a heavy trace metal) as hazardous, meaning that slight exposure can endanger human health. Nine other metals have been defined as hazardous candidates, meaning that they are potential hazards that must be kept under review. They are barium, cadmium, copper, lead, manganese, nickel, tin, vanadium and zinc. Of these, all but manganese are trace metals and all but barium are heavy metals. In selecting these metals, the United States Environmental Protection Agency deliberately ignored metals known to be toxic to humans where the concentrations in the environment, including the industrial environment, are still very low, e.g. antimony and arsenic.

This chapter will outline current toxicological knowledge relating to the metals listed by the United States Environmental Protection

Agency as hazardous or hazardous candidates. In addition, brief mention will be made of the hazards that may be posed by arsenic, chromium, iron and selenium. This is an arbitrary selection, but most, if not all, of the metals presenting quantitatively important environmental risks are included. The amount of space devoted to each metal should not be taken as any measure of its environmental significance. All it reflects is the amount of information that is available.

6.1 Beryllium

Beryllium is released during the burning of coal, but the main environmental hazard is to those working in industries where beryllium is produced or used, e.g. the manufacture of nuclear reactors, aircraft and rockets. In humans, beryllium has been shown to damage skin and mucous membrane. It accumulates in the lungs where it causes beryllium disease (berylliosis). It may cause cancer in lungs and bone marrow. It is not excreted from mammalian tissue and, therefore, its effects are cumulative. At the biochemical level, beryllium competes with magnesium for enzyme sites and has been shown to inhibit DNA polymerase, thymidine kinase and alkaline phosphatase.

6.2 Mercury

6.2.1 Occurrence and use

Mercury is concentrated in various ores, the principal one being cinnabar (HgS). It has been mined since 700 BC and is currently used industrially in three forms: as the metal, in organic compounds and in inorganic compounds. The greatest use of mercury is in the production of electrical apparatus. The second greatest use is in the chloro-alkali industry, which produces chlorine and caustic soda by electrolysis of sodium chloride solution using mercury as the cathode of the electrolysis cell. The third greatest use worldwide is in fungicides (see Section 5.4), including seed dressings, thought it should be noted that these have been banned in a number of countries.

Nearly all the mercury used by man eventually enters the natural environment. To this may be added amounts released during the production of mercury and other metals, from coal burning and from weathering of rocks and degassing from the earth's crust.

6.2.2 Toxicities of different forms
All forms of mercury are potentially toxic but the toxicities vary considerably. The least toxic are the inorganic mercurials. They are

not readily absorbed from the gastro-intestinal tract. Once absorbed they may accumulate in the liver and kidney but normally they are excreted quite rapidly in the urine. It is worth noting that mercury amalgam has long been used in dentistry to fill teeth without any toxic effects being noted, even over most of a human lifetime. Mercury vapour is the most hazardous of the inorganic forms because it can diffuse through the lungs into the blood and then into the brain, where serious damage can occur. Arylmercurials are not much more toxic than the inorganic forms since they are readily broken down to inorganic derivatives in the tissues. Alkylmercurials are the most toxic mercury compounds so far studied. They are fairly stable and have long retention times in the tissues. Therefore, they readily accumulate to high concentrations. Their lipid solubility gives them an affinity for nervous tissue which accounts for many of their harmful effects. In addition, they have been reported to cause abnormalities in cell division and to increase the frequency of chromosome breakages. Some of these abnormalities may be due to combination of mercurials with sulphydryl groups. Inhibition of enzymes in this way has been demonstrated frequently. So far, no effective treatment for mercury poisoning has been developed, though British antilewisite (BAL) or calcium ethylenediaminetetraacetate (Ca_2EDTA) may have some alleviating effect.

6.2.3   Environmental transformations

Unfortunately, in assessing the risk from mercury in a particular environment, it is not enough to know the form in which it entered that environment because various transformations can take place. Probably the most serious of these is the transformation of metallic mercury to methyl and dimethyl derivatives by anaerobic micro-organisms, especially *Clostridium cochlearium*, in aquatic sediments. This may also occur in decaying fish. Under aerobic conditions, this transformation can be brought about by *Pseudomonas* spp. and by the fungus, *Neurospora crassa*. In essence, it represents conversion of the least toxic form of mercury to the most toxic. Other transformations which can be brought about by bacteria are the following: phenyl, ethyl and methyl mercury can be reduced to elemental mercury and benzene, ethane and methane, respectively; phenyl mercuric acetate can be converted aerobically to elemental mercury and diphenyl mercury; mercuric ions can be reduced to elemental mercury. Not all the transformations observed are biological. Under alkaline conditions, methyl mercury converts to the more volatile dimethyl mercury. Under oxidizing conditions, in the presence of ultraviolet light, phenyl mercury, alkoxyalkyl mercury and alkyl mercury may break down to give inorganic mercury. Under anaerobic conditions, mercuric ions may combine with hydrogen sulphide to form poorly-soluble mercuric sulphide. With subsequent aeration, mercuric sulphide can be converted to the soluble sulphate which may then be methylated biologically.

6.2.4 *Effects on organisms other than* Homo sapiens

Most micro-organisms are relatively insensitive to mercury and its derivatives. Nitrogen fixing bacteria in the soil require levels of about 100 ppm before they are adversely affected. Normal soil levels are between 0.0005 and 1 ppm. However, marine and freshwater phytoplankton, especially diatoms, are very sensitive to organomercurial fungicides and as little as 0.001 ppm may reduce their photosynthetic efficiency. Many algae and other plants have the ability to absorb and concentrate mercury from the surrounding environment. Droplets of elemental mercury have been found in chickweed. Such high concentrations may cause mitotic disturbances and kill the plants. Fortunately, most agricultural plants do not seem to absorb much mercury. Animals tend to accumulate mercury through their food. Pike can accumulate a concentration of mercury 3000 times higher than that in the water in which they live. Tuna and swordfish show the same ability. Similar observations have been made on predatory birds. Seed-eating birds accumulate mercury where alkylmercury seed dressings are being used. Much less is accumulated where alkoxylakyl compounds are used, and negligible quantities where inorganic mercury compounds are used.

6.2.5 *Effects on* Homo sapiens

The effects of mercury and its derivatives on *Homo sapiens* deserve special consideration because it was mainly these that caused concern about the effects of heavy metals released into the environment in large amounts. The first serious incident to come to light occurred at Minamata Bay in Japan. In this case, comparatively nontoxic inorganic mercury along with some methyl mercury was released in effluent by a chemical factory using mercuric sulphate catalysts in acetaldehyde production. The effluent entered a river running into Minamata Bay. In the sediments, the inorganic mercury was converted to methyl mercury. This accumulated in shellfish and fish which were eaten by the local inhabitants. In consequence, by 1975, 115 people had died and many were left paralysed for life. Others suffered impairment of vision and hearing and other neurological symptoms. Prenatal poisoning of the foetus was observed even in the absence of symptoms in the mother. Since the Minamata Bay incident, another has occurred around the Agana River, Niigata, Japan. This led to 23 deaths. In both these cases, many domestic animals, fish, shellfish and seabirds were affected.

Mercury poisoning of human beings has also been caused by the consumption of food containing high concentrations derived from alkylmercury agricultural seed treatments used to prevent seed-borne disease. In Iraq, treated seeds, intended for planting, were used to make bread. Thousands were poisoned and hundreds died. In the USA, cattle fed on treated grain were slaughtered for human consumption. Again, severe poisoning resulted. In consequence, a number of countries have now banned the use of alkylmercurial seed treatments. A committee of experts brought together by the Food and Agriculture Organization (FAO) and World Health Organization (WHO) has recommended that the use alkylmercurials should be restricted to stocks of cereal seed used for plant breeding or seed production, and never permitted for treatment of cereal seed for export for the production of food.

6.2.6   Assessment of risk

Despite the major incidents referred to above, it seems fairly certain that the average person is at no great risk from exposure to mercury. The normal dietary intake is well below what is thought to be the tolerable limit of 0.3 mg per person per week, of which not more than 0.2 mg should be in the methylated form, according to WHO. For most people, the chances of appreciable exposure from air, pesticides or pharmaceuticals are very limited. Where there is a risk of occupational exposure, this should be minimized by appropriate precautions and medical screening. With regard to the general environment, there is still a need to know more about the concentration and distribution of mercury, especially with regard to those areas where localized high concentrations do exist. Finally, more knowledge is required of the ways in which mercury moves through the environment, and of its effects on living organisms.

6.3   Barium

The most common naturally occurring form of barium is barium sulphate, sometimes called barite or baryta. Barium derivatives are used as fillers for rubber, linoleum, plastics etc., as pigments in paint, in glass manufacture, in the ceramic industry and in various other industrial applications. These applications constitute the main source of risk, primarily to workers in the appropriate industries. Barium salts are highly toxic on ingestion. They cause vomiting and diarrhoea, which may be associated with stomach, intestinal and kidney haemorrhage. They also affect the central nervous system, causing convulsions and have been implicated as a cause of pneumoconiosis. Despite the known toxicity of barium salts, barium sulphate is used to coat the alimentary tract for X-ray photographs since barium absorbs X-rays strongly and thus increases the contrast. The insolubility of barium sulphate minimizes its toxicity, and makes its use acceptable.

6.4 Cadmium

Cadmium is a normal constituent of soil and water at low concentrations. It is usually mined and extracted from zinc ores, especially zinc sulphide. Industrially, cadmium is used as an anti-friction agent, as a rust proofer and in alloys. It is also used in semiconductors, control rods for nuclear reactors, electroplating bases, PVC manufacture and batteries. In the environment, cadmium is dangerous because many plants and some animals absorb it efficiently and concentrate it within their tissues. Normally, however, retention from food by mammals is low but absorption is increased if the mammals are on a low calcium diet. Once absorbed, cadmium associates with the low molecular weight protein, metallothionein, and accumulates in the kidneys, liver and reproductive organs. Very small doses can cause vomiting, diarrhoea and colitis. Continuous exposure to cadmium causes hypertension, heart enlargement and premature death. There is some evidence suggesting that cadmium can induce chromosome abnormalities and may exert a carcinogenic effect on the lungs.

In 1955 there was a cadmium poisoning incident in Northern Japan caused by the accumulation of cadmium in rice and soya beans. This was characterized by extreme lumbago and skeletal collapse, apparently due to increasing bone porosity caused by inhibition of bone repair mechanisms. This was called Itai-itai (ouch-ouch) disease because of the severe pains associated with it. Apart from this, the principal hazard again is to those working in the industries which use cadmium where appropriate precautions must be taken.

6.5 Copper

Copper is one of the most abundant trace metals. It is widely used in its metallic state, either in the pure form or in alloys. For almost all organisms it is an essential micronutrient. It may occur in very high concentrations in water, sediments and biota in some localized areas as a result of mining activities, of intensive use of copper pellets in pig rearing, or of the application of copper fungicides (see Section 5.4). However, there is no evidence of food chain magnification. Hence, most toxic effects are due to immediate exposure to the element. All organisms are harmed by excessive concentrations, which may be as low as 0.5 ppm for algae. Most fish are killed by a few parts per million. In higher animals brain damage is a characteristic feature of copper poisoning.

6.6   Lead

Lead is widely-distributed naturally but the greatest risks normally arise from the emissions to the environment associated with human use of the metal and its derivatives. Fumes and dust come from the smelting of lead, from the manufacture of insecticides, paint, pottery glazes and storage batteries, and from gasoline containing lead additives. Sewage sludge may contain very high levels of lead and its use as a fertilizer may contaminate soils. High levels may occur in urban air as a result of the high traffic density and associated emission of lead from gasoline additives.

Lead affects micro-organisms by retarding the heterotrophic breakdown of organic matter. Little is known regarding the toxicity of lead to plants, where it tends to be localized in the root system. Animals may take in lead by inhalation or ingestion. Only tetra-ethyl lead can be adsorbed through the intact skin. Absorption is very slow but excretion is even slower, so that lead tends to accumulate. Most of the lead is taken up by red blood cells and circulated throughout the body where it may concentrate initially in the liver and kidneys. Thereafter, it may be redistributed to the bones, teeth and brain. In the bones, the lead is immobilized and does not contribute to immediate toxicity but it is a potential hazard since it may be mobilized during feverish illness, as a result of cortisone treatment, and in old age.

Anaemia is the first symptom of chronic lead poisoning in animals because lead interferes with the synthesis of haem. This is associated with abdominal symptoms, which may include nausea, vomiting and abdominal pain. More serious is the degeneration of tissue in the central nervous system which is also observed, especially in children.

Awareness of the dangers from lead has resulted in measures to reduce the human contribution to the environmental load. In particular, progress is being made towards eliminating lead additives from gasoline. Further, the lead pipes and tanks, which formerly carried much of the supply of drinking water, are being replaced by copper and plastic pipes and tanks. These two measures should ensure that the risk to the majority of the human population becomes negligible. However, there will still be a risk to children who eat flakes of paint containing lead or even chew toothpaste tubes made from lead alloys.

6.7 Manganese

Manganese is of widespread occurrence and is of considerable importance in the manufacture of steel. Biologically, it is an essential micronutrient for most organisms. However, in excessive amounts it affects animals adversely, causing cramps, tremors and hallucinations, manganic pneumonia and renal degeneration.

6.8 Nickel

Nickel is used in various forms for nickel plating, as a catalyst, as a mordant and in ceramic glazes etc. Again it is a micronutrient for most organisms but excessive quantities have toxic effects. In animals these include dermatitis and respiratory disorders, including lung cancer following inhalation. Amongst enzymes inhibited are cytochrome oxidase, isocitrate dehydrogenase and maleic dehydrogenase. A particularly poisonous derivative of nickel is nickel tetracarbonyl.

6.9 Tin

Tin is widespread in occurrence and in use, primarily for making tinplate and various alloys and compounds. It is an essential micronutrient and the main cause for concern regarding its toxicity has been the development of trialkyl-tin and triaryl-tin compounds having powerful biocidal properties. These are used on growing crops as fungicides and insecticides. They are also used as antimicrobial agents and in marine anti-fouling paints. Great care must be taken in their use since they can damage crops and, in animals, accumulate in the central nervous system with harmful effects.

6.10 Vanadium

Vanadium is widely distributed. It is used as an alloying element for steels and irons, in making oxidation catalysts and in colouring agents used in the ceramic industry. Large amounts enter the atmosphere from the burning of some petroleums. It is an essential micronutrient and may be accumulated by some marine organisms to concentrations many times higher than those in the surrounding water. Excessive levels in animals inhibit tissue oxidation and synthesis of cholesterol, phospholipids and other lipids, and amino acids. Such levels may also cause precipitation of serum proteins.

6.11   Zinc

Zinc makes up only 0.004% of the earth's crust. Its most important use is as a protective coating on other metals, particularly in galvanizing iron and steel. It is an essential micronutrient and is generally regarded as one of the less hazardous elements, though its toxicity may be enhanced by the presence of arsenic, lead, cadmium and antimony, as impurities. Toxic effects have been observed from the inhalation of fumes from galvanizing baths. The 'zinc fever' produced is characterized by chills, fever and nausea. Removal from the fumes leads to complete recovery. Zinc chloride fumes have sometimes caused fatal oedema of the lungs. Zinc or galvanized containers are not recommended for food storage but are acceptable for storing drinking water. This is because acidic foods can dissolve enough zinc from the container to cause poisoning. A factor which serves to minimize the risk of zinc poisoning is that it appears to be lost along food chains, unlike methyl mercury or cadmium, for example, which accumulate.

6.12   Arsenic

Arsenic is a metal of wide occurrence. It is used in alloys, pesticides, wood preservatives and some medical preparations. It was formerly used in paint pigments, but this use ceased when it was found that, under damp conditions, moulds converted the arsenic to the highly toxic gases, arsine and trimethyl arsine. Arsenic is a cumulative poison, causing vomiting and abdominal pains prior to death. It may also cause dermatitis and bronchitis, and may be carcinogenic to tissues of the mouth, oesophagus, larynx and bladder. At the cell level, it can uncouple oxidative phosphorylation and compete with phosphorus in metabolic reactions.

Arsenic is concentrated by organisms exposed to it and accumulates along food chains. Accumulation in fish seems to be favoured by increasing salinity. Crabs and lobsters have been especially noted for accumulating high concentrations, but no cases of human poisoning seem to have arisen from this.

6.13 Chromium

Like most of the metals discussed here, chromium is widely distributed. For most organisms, it is essential as a micronutrient in trace quantities for fat and carbohydrate metabolism. In industry, it is used in making steel alloys, in chromium plating and in leather tanning. Chromates are water soluble and can poison sewage treatment processes. The chromium ion can exist in four valency states; Cr^{2+}, Cr^{3+}, Cr^{5+} and Cr^{6+}. Of these, the hexavalent ion is the most toxic and it should be reduced to the trivalent state to form insoluble products before chromium waste is released into the environment. Hexavalent chromium has been implicated in poisoning in Japan. In this case, aerosols from chromium refining plants appear to have affected a considerable number of people, causing lung cancer. Besides this, it has been shown that chromates act as irritants to the eyes, nose and throat, and chronic exposure may lead to liver and kidney damage. A characteristic effect on human beings is the appearance of perforations in the nasal septum. At the cell level it appears that hexavalent chromium may cause chromosome abnormalities. Chromium is particularly dangerous because it accumulates in many organisms. Some aquatic algae have been shown to concentrate it 4000 times above the level of their immediate environment.

6.14 Iron

Iron is the fourth most abundant element in the earth's crust. Its greatest use is for structural iron and steel, but it is also used for making dyes and abrasives. It is an essential micronutrient in trace quantities for most organisms, but ingestion of excessive amounts may result in the inhibition of activity of many enzymes. The amounts consumed must be very large because only a small proportion of all iron ingested is absorbed from the gastro-intestinal tract. Inhalation of iron dust can cause benign pneumoconiosis and can enhance harmful effects of sulphur-dioxide and various carcinogens.

Many streams are poisoned by high levels of iron in acid mine drainage. Pyrite, iron sulphide, is often found in close association with coal deposits. Upon exposure to moisture and atmospheric oxygen, the ferrous iron is oxidized to the ferric state, a reaction which is frequently accelerated by bacteria of the Thiobacillus – Ferrobacillus group. The ferric iron can then react with sulphide in the presence of water to produce sulphuric acid, or react directly with water to produce a yellow, flocculent mass of ferric hydroxide. Besides being acidic, water affected in this way becomes deficient in oxygen. Such poisoning of streams is reckoned to be one of the main causes of fish kill in the United States. Although particularly associated with mining, streams running through iron-laden strata may become poisoned spontaneously.

6.15   Selenium

Strictly selenium is not a metal, though it has certain metallic properties. It is a member of the sulphur-group, produced as a byproduct of the extraction or copper, nickel, gold and silver ores. It is used in electronics, and in paints and rubber compounds. for most organisms it is an essential micronutrient but it can be toxic at very low concentrations. The maximum permissible concentration in drinking water is 0.01 ppm. Poisoning of livestock has occurred where cattle have eaten plants of the Brassica family which have taken in selenium and incorporated it in cysteine and methionine in place of sulphur. Selenium itself irritates the eyes, nose, throat and respiratory tract. It can cause cancer of the liver, pneumonia, liver and kidney degeneration and gastro-intestinal disturbances.

7 Atmospheric Toxicants

Many substances in the atmosphere may be toxic, including many already discussed, but in this chapter coverage will be restricted almost entirely to those that are commonly regarded as air pollutants, i.e. carbon monoxide, hydrocarbons, sulphur oxides, particulates and nitrogen oxides. This list is in order of total mass emitted per annum from all sources in the USA in 1970, and the order is likely to be similar in other industrialized counties. However, not all these toxicants are equally harmful and, in assessing the risks they pose, allowance must be made for their varying degrees of toxicity. An attempt has been made to do this by assigning effect factors to the toxicants mentioned. The effect factor is defined as the ratio of toxicant LC_{50} to carbon monoxide LC_{50}. Total toxicant emitted to the atmosphere is multiplied by its effect factor to give a new datum, the emission effect. In terms of emission effects, the hydrocarbons are the most serious atmospheric toxicants.

The atmospheric toxicants considered in this chapter are largely products of the internal combustion engine. In the USA in 1970, 54.5% by mass of these toxicants came from vehicles, 16.9% from stationary fuel combustion, 13.7% from industrial processes and 4.2% from solid waste disposal.

Attempts to regulate the emission of atmospheric toxicants have necessitated the determination of safe atmospheric concentrations. In relation to the industrial environment, these may be established by reference to threshold limiting values (TLVs), formerly known as maximum allowable concentrations, (MACs). The threshold limiting value of a toxicant, or mixture of toxicants, is defined as the maximum concentration to which it is believed healthy workers may be repeatedly exposed without ill effect, on the basis of an eight hour working day. In relation to the atmosphere at large, the threshold limiting value is replaced as a reference point by the 'three minute mean concentration'. This is defined as the maximum permissible concentration of a toxicant, or mixture of toxicants, averaged over three minutes, under the conditions most likely to favour high concentrations at ground level beside a stationary source. The three minute mean concentration is often calculated as a fraction of the threshold limiting value, usually one-thirtieth or one-fortieth. Where there is a risk of accumulation of a toxicant, e.g. a heavy metal, an absolute mass emission limit may be established. The absolute mass emmission limit is the total amount of toxicant that may be emitted from a stationary source per hour, and which may not legally be exceeded.

In this chapter, atmospheric toxicants will be considered individually under separate headings but their interactions will not be ignored, since these interactions are often the direct cause of harmful effects. In addition to the toxicants already mentioned as being quantitatively important, fluorides and asbestos dust will be considered because their toxic effects, though localized, have caused particular concern.

7.1 Carbon Monoxide

7.1.1 Emission to and removal from the atmosphere

Carbon monoxide is the most abundant and widely distributed of the toxicants described in this chapter. It is a colourless, odourless and tasteless gas, slightly lighter than air and only very slightly soluble in water. It can burn but does not itself support combustion.

On an annual basis, about ten times as much carbon monoxide enters the atmosphere from natural sources as from transportation and other human activities. Most of this (77.6%) comes from atmospheric oxidation of methane arising from the breakdown of organic residues, especially in tropical regions. Smaller contributions come from the oceans (3.9%) and from the growth and breakdown of chlorophyll in plants (2.6%). This naturally generated carbon monoxide does not pose any serious problems because the rate of emission at any given place is relatively low. However, manmade carbon monoxide tends to be released rapidly in urban areas. This results in localized concentrations 50 to 100 times higher than the global average.

The concentration of carbon monoxide in the atmosphere at any time depends not only on the rate of production but also on the rate of removal. This occurs mainly in the soil, though a small amount is oxidized to carbon dioxide in the lower atmosphere. In the soil, fourteen species of fungi have been identified as the active agents oxidizing carbon monoxide to carbon dioxide. The activity of these fungi varies with the type of soil, being least active in desert soils and most active in the tropics. Cultivated soils are less satisfactory to these fungi than are those covered with natural vegetation. The total capacity of soil fungi for carbon monoxide oxidation seems to be more than adequate to deal with the carbon monoxide produced on a global basis. Unfortunatley, the urban areas, where most anthropogenic carbon monoxide is released, have among the poorest soil reservoirs of these fungi which, therefore, cannot make an effective contribution to lowering the localized high concentrations in these areas.

7.1.2 Toxic effects

Carbon monoxide has little effect on plants and micro-organisms at the maximum concentrations (about 15 ppm) which normally occur, even in urban areas. However, high concentrations (100 ppm and above) are lethal to many animals. The carbon monoxide combines with haemoglobin forming carboxyhaemoglobin, thus reducing the ability of the blood to carry oxygen. There is no doubt that levels of carboxyhaemoglobin above 5% of the total haemoglobin have harmful effects, ranging from headache, fatigue and drowsiness, to coma, respiratory failure and death. It is probable that levels as low as 1% can adversely affect physiological function. For human beings, the equilibrium percentage of carboxyhaemoglobin in the blood during continuous exposure to an ambient-air carbon monoxide concentration of less than 100 ppm can be estimated from the following equation:

$$\%COHb \text{ in blood} = 0.16 \times \text{concentration of CO in air (ppm)} + 0.5$$

where COHb = carboxyhaemoglobin
0.5 = normal background percentage COHb in blood.

Thus, the blood concentration is directly related to the ambient carbon monoxide concentration. The rate of equilibration varies with the physical activity of the exposed person, increasing as this becomes more strenuous. As far as human beings are concerned, allowance must also be made for voluntary intake of carbon monoxide from smoking. This increases the background blood content of carboxyhaemoglobin 2 to 4 times above the level in non-smokers.

7.1.3 Control

The main source of anthropogenic carbon monoxide is the internal combustion engine and so it has become the prime target for control procedures. Since other toxicants are simultaneously released, there is always the possibility that a method of lowering the concentration of carbon monoxide may increase the concentration of these. For example, supplying a stoichiometric air/fuel mixture to the internal combustion engine will give low carbon monoxide and hydrocarbon emissions, but high emissions of nitrogen oxides. The principal technique evolved to reduce carbon monoxide emission has been the development of exhaust system reactors, which convert carbon monoxide to carbon dioxide, and hydrocarbons to carbon dioxide and water. There are two types of reactor: those that are catalytic and depend on platinum or palladium, or a mixture of both, and those that use a high temperature chamber. None of the catalysts available is entirely satisfactory. They are all poisoned by lead and sulphur compounds which may be present in gasoline, and by ethylene dibromide which is added to leaded gasolines to prevent build-up of lead deposits in engines. They are also poisoned by organothiophosphate compounds released from lubricating oil. It is even possible that platinum and palladium released from catalytic reactors into the environment may prove to be toxic. Certainly, water soluble platinum salts have harmful effects. High temperature chambers do not have any of these shortcomings but they are more expensive to operate because they increase fuel consumption. Other approaches to reducing carbon monoxide emission include improvements in engine design, development of substitute fuels and development of new power sources. None of these has so far proved entirely satisfactory.

7.2 Hydrocarbons and Photochemical Derivatives

7.2.1 *Emission to and removal from the atmosphere*

Hydrocarbons can exist as gases, liquids or solids under normal environmental conditions. Under such conditions, molecules containing four carbon atoms or less are gases, while those with five or more are liquids or solids. Most of the hydrocarbons contributing to air pollution have twelve or less carbon atoms per molecule and, therefore, are either gases or volatile liquids. These hydrocarbons may be either aliphatic or aromatic.

As with carbon monoxide, most hydrocarbons enter the atmosphere from natural sources. Methane, the simplest hydrocarbon, makes the largest contribution on a global basis. It is produced mostly by bacterial decomposition of organic matter, particularly in swamps and marshes. Plants emit the more complex hydrocarbons, terpenes and hemiterpenes, the total production of which is about half that of methane. Human activities contribute only about 15% of the total annual release of hydrocarbons to the atmosphere but, as with carbon monoxide, this release, mostly from transportation, is highly localized in urban areas.

The toxic effects of hydrocarbons in the atmosphere are mainly due to derivatives produced by photochemical oxidation (Fig. 7.1).

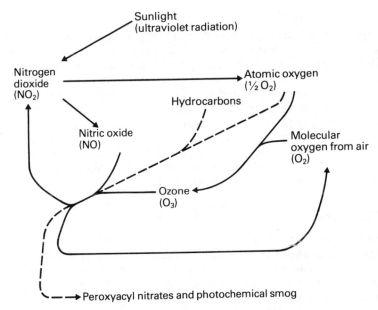

Fig. 7.1 The nitrogen dioxide photolytic cycle, showing the interaction of hydrocarbons to produce photolytic smog.

Such toxic derivatives are sometimes referred to as secondary pollutants. Photochemical oxidation frequently involves ozone and peroxyacyl nitrates, which have the general formula RCO_3NO_2 where R is a hydrocarbon group. Ozone is produced in the atmosphere by the action of sunlight on nitrogen dioxide (see Section 7.5.1), which splits it into nitric oxide and atomic oxygen. Each atom of oxygen can then react with a molecule of oxygen to give ozone. If hydrocarbons are present, they can react with the ozone to form hydrocarbon free radicals which are very reactive because of the unpaired electrons which they contain. These free radicals can react with each other, with atmospheric toxicants and with other components of air to produce photochemical smog. Hydrocarbons vary widely in their potential to form free radicals. Therefore, the nature of a hydrocarbon mixture must be known before its contribution to photochemical smog can be assessed. In general, unsaturated compounds are the most reactive. Following reactions with ozone, atomic oxygen, molecular oxygen and nitrogen dioxide, they give rise to the group of toxicants known as peroxyacyl nitrates, to which reference has already been made. These toxicants, like ozone, can react further with hydrocarbons to generate free radicals.

7.2.2 Toxic effects

Photochemical smog harms both plants and animals. The damage has been attributed mostly to ozone and peroxyacyl nitrates. Both kill leaf tissue. Citrus, forage and salad crops, and coniferous trees seem especially sensitive. Peroxyacyl nitrates in photochemical smog can cause eye irritation in animals. On the other hand, ozone, at the maximum levels so far detected in the lower atmosphere (about 0.1 ppm), has not been shown to have any toxic effects. However, bearing in mind that low concentrations of ozone have been shown to cause degradation of organic polymers such as rubber and cellulose, it seems likely that some harmful effects do occur and will eventually be detected.

7.2.3 Control

Control of hydrocarbon emission from transportation involves the same treatment of the exhaust fumes as does removal of carbon monoxide. In addition, loss of hydrocarbons by evaporation from the fuel tank and carburettor is reduced by installation of a collection system which eventually returns them to the fuel induction system. Hydrocarbon emissions from stationary sources are controlled by use of afterburning, adsorption, absorption and condensation. Afterburning completes the oxidation of the hydrocarbons to carbon dioxide and water. For adsorption, a bed of activated carbon is used. This bed is periodically cleaned with steam, from which the hydrocarbons may be recovered following condensation. Where possible, the recovered hydrocarbons may be utilized for other processes. Otherwise, they may be disposed of by oxidation to carbon dioxide and water. Absorption is achieved by passing exhaust fumes through a liquid in which the hydrocarbons will dissolve or become suspended. Again, they may be recovered for subsequent use or they may be eliminated by oxidation.

7.3 Sulphur Oxides

7.3.1 Emission to and removal from the atmosphere

The sulphur oxides are sulphur dioxide and sulphur trioxide, usually emitted in a mass ratio of about 100 to 1. Sulphur dioxide is a colourless, nonflammable gas under normal environmental conditions, and has an irritating smell at concentrations above 3 ppm. Sulphur trioxide is not normally found in the atmosphere because it reacts rapidly with water to form sulphuric acid. The sulphur oxides released to the atmosphere come almost entirely from human activities, especially coal burning, though small amounts come from burning of fuel oil and smelting of sulphide ores. More are produced in the atmosphere by the oxidation of hydrogen sulphide. This contributes up to 57% of the total production of all sulphur oxides. The prime source of atmospheric hydrogen sulphide is the decay of organic matter. Industrial activities and volcanic activity make small contributions. The main oxidant involved may well be ozone.

Much of the atmospheric sulphur dioxide is ultimately oxidized to sulphur trioxide, which is converted to sulphuric acid as described above. The sulphuric acid then forms sulphates which settle out of the atmosphere or are washed out by rain. The oxidation of sulphur dioxide in the atmosphere is rapid. Catalytic oxidation may occur in water droplets or on the surface of solid particles. In water droplets, the oxidation is thought to involve molecular oxygen, with iron and manganese salts as catalysts. The iron and manganese salts are largely derived from the fly ash of burning coal. The ash particles may also act as nucleation sites for water droplet formation. In addition, photochemical oxidation may occur with ozone or peroxyacyl nitrates in photochemical smog. The result of either of these processes is a mist of sulphuric acid droplets. Catalytic oxidation stops when the sulphuric acid concentration in the droplets reaches one molar, possibly because of the resultant reduction in sulphur dioxide solubility. Neutralization of the acid by metal oxides or ammonia increases the solubility of sulphur dioxide and permits catalytic oxidation to resume. Ammonia is probably particularly important in this process because it occurs naturally in the atmosphere. The solubility of ammonia increases with decreasing temperature and, therefore, its effect may be especially marked at high altitudes where temperatures are low.

7.3.2 Toxic effects

Chronic exposure of plants to low concentrations (about 0.01 ppm) of sulphur dioxide causes leaf yellowing as a result of inhibition of chlorophyll synthesis. This is associated with accumulation of sulphate in the leaves. High concentrations of sulphate (about 0.5 ppm and above) cause rapid leaf destruction. As always, response varies from species to species and with environmental conditions, most notably the presence of other pollutants. For example, ozone and nitrogen dioxide have been shown to exert synergistic effects with sulphur dioxide. The sulphuric acid mists associated with high sulphur dioxide concentrations in the atmosphere are another major source of damage, again primarily through effects on leaves.

Acute sulphur dioxide effects on animals are mainly related to the respiratory system. A concentration of 1.6 ppm will cause reversible bronchiolar constriction, and detectable respiratory effects increase above this level. Below 25 ppm, irritant effects occur mainly in the upper respiratory tract and the eyes, because the high water solubility of sulphur dioxide ensures that it dissolves in the first tissue water with which it comes in contact. Studies of the effects of chronic exposure to low levels of sulphur dioxide show an increased incidence of respiratory infection. This must be a particular hazard for those already suffering from respiratory deficiencies. Sulphur dioxide dissolved in tissue water is converted oxidatively to sulphuric acid and sulphates as previously described. Sulphates are much more powerful irritants than sulphur dioxide itself and the problem is enhanced by the presence of sulphate aerosols (see Section 7.4) in the atmosphere wherever sulphur dioxide is produced. Concentrations of sulphate as low as 0.002 ppm have been shown to have harmful effects on susceptible people, e.g. asthmatics, and it is not uncommon for concentrations in the atmosphere of this order to be reached or exceeded. An unfortunate consequence of the introduction of catalytic units in vehicle exhaust systems, to remove carbon monoxide and hydrocarbons, has been an increased emission of sulphate, since the catalysts convert the trace amounts of sulphur oxides present into sulphuric acid. Though the total produced is not great, it may well be that the hazard from increased sulphate emissions exceeds the benefits from reduced carbon monoxide and hydrocarbon levels.

7.3.3   Control

Modifying the catalytic units in car exhaust systems to reduce sulphuric acid production may involve changing the method of catalyst use, adding a sulphate trap, or completely removing sulphur from gasoline prior to its use. As far as coal burning, the main source of anthropogenic sulphur oxides, is concerned, there are three possible ways of reducing emission. Firstly, one can use coal with a very low sulphur content. Unfortunately, there are only a few regions where this is economically feasible. Secondly, sulphur can be removed from coal by a water washing process. However, this is expensive and removes only pyritic sylphur. Thirdly, coal can be gasified and the sulphur oxides produced trapped by a suitable chemical reaction. This method looks the most promising, but it is still fairly costly in operation.

7.4 Particulates

7.4.1 Emission to and removal from the atmosphere

Particulates is a term used to cover all small, solid particles and liquid droplets in the air, except droplets of pure water. An alternative term is aerosols. Other terms may be used to define different types of particulate. Mists consist of suspended liquid droplets; smoke is very small soot particles produced by burning; fumes are condensed vapours; dusts are fine particles produced by breakdown of solids. Particulates may also be divided into those that are viable, e.g. bacteria, fungi, moulds and their spores, and those that are nonviable. Viable particulates pose special problems. These are best considered in depth in a microbiology textbook. Hence, the following discussion will relate only to the nonviable particulates.

Nonviable particulates may be formed by the breakdown of material or by agglomeration. Sea-salt aerosols, by far the greatest single source of atmospheric particulates, are formed by the bursting of seawater air bubbles which releases many tiny droplets of seawater. The water quickly evaporates leaving the sea-salt particles in the atmosphere. Other processes producing particles by breakdown are soil erosion leading to dust storms, volcanic activity and combustion. Particulates released directly into the atmosphere by any of above processes are called primary particulates. Secondary particulates are those formed in the atmosphere by agglomeration, e.g. by the reactions of gases followed by subsequent solution of the products in water droplets. The production of sulphuric acid from sulphur oxides is an example of this. In total, such secondary particulates slightly exceed in quantity the amount of sea-salt aerosol entering the atmosphere each year. Secondary particulates constitute the main anthropogenic contribution to particulate pollution but amount to only 20% of the yearly natural production of secondary particulates. Similarly, anthropogenic production of primary particulates amounts to less than 8% of the natural production on an annual basis. The main source of anthropogenic particulates is generally thought to be industry, followed by domestic burning of coal and other fuels. However, this view is based on the assumption that soil erosion is a natural process. When one considers that erosion has frequently been due to bad agricultural practice such a view is not tenable and it may well be that, on a global basis, agriculturally generated soil dust is the principal anthropogenic particulate pollutant.

Particulates inevitably vary enormously in their chemical

composition. Soil dust contains largely calcium, aluminium and silicon compounds. Smoke may contain many organic compounds. Secondary particulates frequently contain ammonium sulphate or ammonium nitrate. Some particulates contain pesticide residues, others contain heavy metals. Thus there are hazards quite distinct from the particulate nature of the carrier.

Particulates range in diameter from 0.2 to 5000 nm, but those of diameter less than 100 nm behave like molecules and generally aggregate. Particles of diameter larger than 1000 nm sediment out of the atmosphere under gravity. Thus, most particles in the atmosphere at any time are of diameter between 100 and 1000 nm. Eventually all atmospheric particles reach the earth's surface. Under dry conditions, this will occur by sedimentation, impaction or diffusion. Impaction refers to particles forcibly deposited under the action of wind. Diffusion refers to the normal random motion of small particles in a gas. As a result of this some collide with and adhere to the earth's surface. However, the deposition of most particulates is associated with rain or snow, i.e. wet precipitation. Particles may form nuclei for water condensation and be carried down in the resultant rain, i.e. rainout. Alternatively falling rain or snow will collect particulates as it passes through the atmosphere, i.e. washout. One consequence of increasing levels of sulphur oxides and nitrogen oxides in the atmosphere has been that wet precipitation has become measurably more acidic, especially in industrialized regions. Increased fish mortality in these areas has been attributed to this. Another result may be increased leaching of soil nutrients causing loss of soil fertility.

7.4.2 Toxic effects

Little is known of the effects of particulates on plants. These are likely to be variable, since the particulates vary so much in chemical composition. However, one general effect will be the reduction of photosynthesis by prevention of light reaching the leaves and by interference with carbon dioxide uptake.

The most marked effects on animals involve the respiratory system. In humans, particles larger than 5000 nm in diameter do not pass beyond the upper respiratory tract. Particles of between 500 and 5000 nm diameter may reach the bronchioles. However, these are removed by ciliary action to the pharynx where they are mostly eliminated through the gastro-intestinal tract by swallowing. Particles less than 500 nm in diameter may reach the alveoli. Such particles may stay there for periods of up to several years since alveolar membranes have no cilia. If particles do penetrate the respiratory tract, they may exert various harmful effects, e.g.

slowing the ciliary beat and inhibiting the removal of harmful substances in the mucous flow, and thus cause such illnesses as bronchitis (Fig. 7.2). Furthermore, they may be carrying toxic substances, either as minor or major components. In fact, the range of possible toxicities is almost without limit and made more difficult to assess by the potential for synergistic and antagonistic effects.

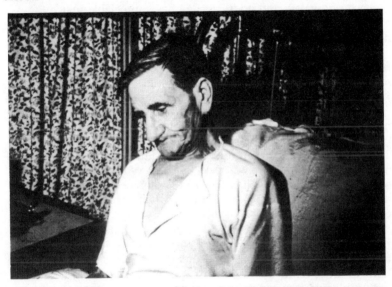

Fig. 7.2 A bronchitic patient. Bronchitis may be caused by atmospheric toxicants and is made worse by them (see Appendix 2) (from the Diana Wyllie slide set 'The Air you Breathe').

7.4.3 Control

Attempts to reduce anthropogenic release of particulates must include improved agricultural practice. However, most attention has been paid to industry and four processes have been developed to lower the industrial contribution. The gravity settling chamber is effective in removing particles of diameter greater than 50 000 nm from gases. The cyclone collector uses centrifugal force to remove particles with diameters 5000–20 000 nm. Wet scrubbers remove particulates, as well as liquids and gases, with variable efficiency, depending on the design. Finally, electrostatic precipitators are often an effective way of removing primary particulates.

7.5   Nitrogen Oxides

7.5.1   *Emission to and removal from the atmosphere*

The nitrogen oxides found in the atmosphere are nitrous oxide, nitric oxide and nitrogen dioxide. Nitrous oxide is colourless, with a slightly sweet taste and smell. Its toxic effects are minimal. Nitric oxide is colourless and odorless, and has appreciable toxicity. Nitrogen dioxide is a reddish-brown gas with a choking smell. Like nitric oxide, it has considerable toxicity.

In general, more nitrogen oxides come from natural sources than from human sources. Almost all nitrous oxide comes from natural sources, which account for 80% of the atmospheric content. However, nearly all the nitrogen dioxide is anthropogenic.

Microbial decomposition of organic compounds in the soil is the main source of both nitrous oxide and nitric oxide. Burning of fossil fuels is the principal source of anthropogenic nitrogen oxides. The ratio of nitric oxide to nitrogen dioxide emitted varies with the nature of the combustion process, but the total amount of nitric oxide is always greater than that of the dioxide. In the United States, transportation accounts for over half the total anthropogenic production of nitrogen oxides.

In the atmosphere, nitric oxide and nitrogen dioxide become involved in the nitrogen dioxide photolytic cycle (Fig. 7.1.). Firstly, nitrogen dioxide molecules split under the influence of ultraviolet light to nitric oxide and atomic oxygen. The atomic oxygen reacts with molecular oxygen to form ozone. The ozone reacts with nitric oxide to give nitrogen dioxide and molecular oxygen. Thus the cycle is completed. As outlined this cycle produces no net change but if hydrocarbons are present, as is usually the case, they interact with nitrogen oxides and ozone and unbalance the cycle. This leads to accumulation of various toxicants, including the peroxyacyl nitrates (see Section 7.2.2.).

Whether or not nitrogen oxides are removed from the atmosphere by soil micro-organisms is still unclear. However, there is no doubt that their removal from the atmosphere involves conversion to nitric acid, possibly by a reaction with ozone. This nitric acid is carried to the earth's surface as nitrate salts in rainfall or dust.

7.5.2   *Toxic effects*

Damage to plants from high levels of nitrogen dioxide in the air has been observed near factories producing nitric acid. In laboratory experiments, nitric oxide has been shown to interfere with photosynthesis in beans and tomatoes, while nitrogen dioxide has been shown to cause necrosis in leaves of cotton, pinto bean and endive plants. These effects were observed at atmospheric concentrations of between 1 and 10 ppm, i.e. higher than the normal concentration which rarely exceeds 1 ppm for nitric oxide or 0.25 ppm for nitrogen dioxide even in the worst affected areas.

Studies of acute lethal toxicity in animals show that nitrogen dioxide is about four times more toxic than nitric oxide. Further, there is little evidence that the concentrations of nitric oxide found in the atmosphere constitute a health hazard. Even proven effects of nitrogen dioxide are associated with concentrations higher than those so far detected in the atmosphere. Such effects are, as is frequently the case with atmospheric toxicants, largely localized in the respiratory tract. Nasal and eye irritation is followed by increasing difficulty in breathing, pulmonary oedema and death. Although nitrogen dioxide may never reach a high enough atmospheric concentration for such effects to be detectable, lower concentrations may be sufficient to sensitize organisms to other noxious compounds or to affect certain sensitive individuals. Again, more research in this area is required.

7.5.3 Control
Most of the methods employed to reduce anthropogenic emission of nitrogen oxides involve modification of the combustion processes involved. Reduced peak temperature, and/or reduced oxygen availability at this temperature, lowers nitrogen oxide levels and this is the main approach used. Catalytic modification of exhaust fumes is difficult because it must be reductive, whereas removal of carbon monoxide and hydrocarbons is oxidative.

7.6 Fluorides

Fluoride particles are released into the atmosphere by brick factories, aluminium smelters and phosphate works. These particles settle on surrounding pasture and there are many recorded observations of cattle losing teeth and suffering lameness from bone malformations following ingestion of fluoride during grazing. This type of poisoning has been referred to as fluorosis. Plants may also suffer from fluoride poisoning. Gladioli of the variety 'Snow Princess' are particularly sensitive. The leaves mottle and turn brown on exposure to concentrations of hydrogen fluoride of $0.5\ \mu g\ m^{-3}$ or more. There are reports of damage to maize and citrus plants at concentrations as low as $4\ \mu g\ m^{-3}$ in the atmosphere. Since atmospheric fluoride concentrations of $1\ \mu g\ m^{-3}$ can lead to plant fluoride concentrations in excess of 30 ppm, it will be seen that chronic exposure to low levels of fluoride may have harmful effects because of fluoride accumulation. Dairy cattle may be particularly at risk.

Deliberate fluoridation of drinking water by public health authorities, with the intention of preventing tooth decay has been a matter or some controversy. Fluoride, as described above, can be very toxic but, at the time of writing, no harmful effects of fluoride have been conclusively demonstrated at the concentrations present in drinking water. Possible correlations with increased incidence of disease have been suggested, but other explanations of the statistical evidence are possible. However, the question remains as to whether the benefit of reduced tooth decay is sufficient to warrant even the remotest possibility of an increase in other, more serious illnesses.

7.7 Asbestos Dust

Asbestos dust is an air pollutant which constitutes a considerable localized hazard of the manmade environment. Asbestos is a fibrous material made up of magnesium silicate. It is used in building and in engine gaskets and brake linings. Its main value lies in its resistance to fire, heat and chemical attack. There are two main forms of asbestos, blue asbestos (crocidolite) and white asbestos (amosite, chrysotile, tremolite). Blue asbestos dust is particularly dangerous but dust from both types of asbestos can have serious effects following inhalation. These effects include lung scarring (asbestosis) and cancer of the bronchii, pleura and peritoneum. Although threshold limiting values for asbestos dust in the atmosphere have been established, it seems likely that no concentration can be regarded as completely safe.

7.8 Temperature Inversions and Other Environmental Phenomena Related to Atmospheric Toxicants

7.8.1 Temperature inversions

Temperature inversions are mentioned here because they can prevent dispersal of atmospheric toxicants from the site of emission. Generally toxicants can be dispersed by wind action but mountains surrounding a valley can hinder this process, as can buildings in a city. Dispersion then becomes dependent on vertical movement of air which, in turn, depends on convection. Air immediately above the earth's surface is warmed by the earth, expands, becomes less dense than the cooler air above it and rises, allowing the cooler air to move in to replace it. In this way, air currents are produced which can disperse toxicants. Sometimes weather conditions will place a cold air mass below a warm air mass so that convection ceases for a considerable period of time, even for several days. This is the phenomenon known as temperature inversion. Then toxicants are trapped and may accumulate in the cold air mass. The warm air mass is referred to as the inversion layer. It is usually cloudless and can transmit sunlight readily, causing photochemical reactions amongst the trapped pollutants and, therefore, high smog concentrations.

7.8.2 The ozone layer

Besides the concern about direct effects of atmosphere toxicants, there has also been concern about indirect effects. For instance, there has been concern about the stratospheric ozone layer which screens out more than 99% of solar ultraviolet rays. These ultraviolet rays constitute high energy radiation, which could cause serious harm to living organisms (see Section 9.3.1). Suggestions have been made that nitrogen oxides from nuclear explosions or from the jet engines of high flying aircraft, and atomic chlorine from the freons used as propellants in aerosol cans, may cause serious depletion of the ozone layer and hence threaten life. The effects of atmospheric nuclear explosions have not been properly assessed, but the risk from jet engines, according to a report of the US Department of Transportation, appears to be minimal. The build up of freons in the atmosphere, however, may pose a substantial risk and steps are being taken to curtail their use.

7.8.3 The 'greenhouse effect' of carbon dioxide

It has been suggested that increased concentrations of carbon dioxide in the air, resulting from human activity, may increase the surface temperature of the earth by reducing heat loss. This follows because carbon dioxide absorbs long wavelength infrared radiation emitted by the earth's surface and, in this way, the temperature of the atmosphere is increased. However, while atmospheric carbon dioxide has been increasing, the temperature of the earth has decreased since 1940. This may either be due to increased cloud cover or to increased atmospheric particulate concentrations, both of which would reduce the amount of radiation reaching earth. Nevertheless, the potential danger of the 'greenhouse effect' should not be ignored, as it is possible that reduction in the levels of the toxicants discussed in this chapter may leave it as the final atmospheric aberration to be corrected.

8 Petroleum and Radionuclides

Toxicants are generally considered to be substances which cause harm to living organisms as a result of some chemical action. Petroleum and radionuclides may be toxicants in this sense, but much of the environmental damage they cause is due to their physical properties. In the case of petroleum, the most obvious effects are due to its hydrophobic nature and its impermeability to oxygen. Thus, when it coats organisms, they die from asphyxiation. Similarly, contaminated areas of the environment suffer from oxygen depletion. The hydrophobic properties are associated with chemical stability and this results in persistence. In the case of radionuclides, most harm is caused by the ionizing radiation which they emit. This radiation modifies or inactivates biological molecules, frequently with fatal results. Like petroleum, radionuclides may be persistent. In both cases, therefore, environmental damage may occur long after imput to the environment has ceased. For this reason special precautions must be taken to minimize the anthropogenic contribution.

8.1 Petroleum and Related Compounds

8.1.1. Petroleum (oil)

Petroleum, generally referred to simply as oil, is a mixture of alkanes containing, on average, about 1% by weight of aromatic hydrocarbons. It has become a matter of serious environmental concern because its extraction and use as an energy source by humans has led to its widespread distribution in the biosphere. This has been most obvious following large-scale marine spillage, either from tanker wrecks or from oil-well blowouts, but the cumulative effects of the many minute spills that have occurred both on land and sea may be at least as serious.

The harmful effects of oil on living organisms may be divided into those that are primarily physical and those that are primarily chemical. Physical effects are caused by oil coating the organisms or their immediate environment. This is very clearly seen when water birds become covered with oil. By matting the feathers, the oil destroys their insulative capacity, reduces buoyancy in the water and prevents flight. In other organisms, oil coating may cause death by asphyxiation. Oil films on the surface of natural waters reduce light transmission and, hence, photosynthetic primary production. Such films also retard oxygen uptake by water and so cause a lower dissolved oxygen concentration and the death of many organisms.

Chemical effects of oil can be related to the components involved. Low boiling point saturated hydrocarbons, at least up to octane, can produce anaesthesia and narcosis in many lower animals. Low boiling point aromatic hydrocarbons are even more toxic, and their greater water solubility tends to enhance their distribution and uptake by aquatic organisms. Benzene, toluene, naphthalene and phenanthrene are amongst the compounds in this group. Benzene characteristically inhibits blood cell formation in bone marrow. All cause local irritation of the respiratory system and excitation or depression of the central nervous system. Many may be mutagenic, carcinogenic or teratogenic. Polycyclic aromatic hydrocarbons are especially dangerous in this respect (see Section 8.1.3). High boiling point saturated and aromatic hydrocarbons

Fig. 8.1 Dispersion of petroleum on beaches by detergent spray. The solvents and detergents in such sprays are both toxic. Where possible oil should be removed by physical means such as skimming off the upper layer of sand (from the Diana Wyllie slide set 'Water Pollution').

may not exert much direct toxicity but may interfere with the responses of aquatic organisms to chemical stimuli, e.g. sex attractants, with equally serious consequences.

Since many of the components of oil are chemically stable and not readily metabolized or excreted once absorbed, they are subject to food chain amplification (see Section 9.2.1). In this way, they may give unpleasant flavours to food for human consumption or render it toxic. It is even conceivable that oil in natural waters may concentrate fat soluble toxicants by a partitioning process, thus enhancing their toxicity.

Not only are there hazards associated directly with oil spills, there are others associated with oil clearance from natural waters and from beaches. Containment with booms followed by removal by skimming devices causes the least damage but is frequently impossible. As an alternative, detergents may be used to disperse the oil (Fig. 8.1). This may give aesthetically satisfactory results but may enhance the biological destruction done by the spill. All detergents have appreciable toxicity and can facilitate uptake of oil by aquatic organisms (see Section 8.1.2). Further, they may cause oil on beaches to penetrate the sand more deeply, thereby prolonging harmful effects on intertidal animals and plants.

8.1.2 Detergents

The detergents considered here are those which include in their formulation synthetic surfactants. The most commonly used surfactants are the linear alkyl sulphonates, e.g. sodium dodecylbenzenesulphonate. These compounds are rapidly degraded by bacteria and do not normally constitute a serious environmental hazard, unlike the branched chain alkyl compounds, e.g. tetrapropylene benzene sulphonate, which they replaced and which were resistant to bacterial degradation. However, all detergents have appreciable toxicity and, if released in large amounts, may reach concentrations which kill many organisms. This is most likely to occur when they are being used as dispersants to break up oil slicks on beaches or in natural waters. Even if the detergents do not kill the organisms directly, they may damage cell membranes sufficiently to allow other toxicants to act. Thus, the benefits of rapid oil dispersal by detergents must be weighed very carefully against the risks involved before their use is approved.

Detergents used for cleaning purposes contain compounds called builders. These chemicals sequester divalent ions such as calcium and magnesium so that they do not interfere with the cleaning action of the surfactant. Builders also make wash water alkaline, thus improving dirt removal. The most common builders used are polyphosphates. They hydrolyse to phosphate when released into the natural environment, and this may contribute to excessive eutrophication (see Section 9.2.3). However, substitutes for polyphosphates have all got drawbacks and it seems likely that their use will continue, but with care being taken to remove them from waste water before it is released. This can be done quite readily by precipitation with aluminium sulphate.

8.1.3 *Polycyclic (polynuclear) aromatic hydrocarbons*
These compounds, e.g. 3,4-benzpyrene and 1,2-benzanthracene (see Table 1.1) have attracted considerable attention. They are believed to be the main carcinogens in cigarette smoke and are potent mutagens. They are found in petroleum and in industrial products, and are also synthesized by plants. They cause mutations either by direct covalent bonding with DNA or by wedging themselves into the DNA helix (intercalation).

Much more needs to be known about the natural background level of these compounds and the extent to which organisms have evolved metabolic systems, e.g. the cytochrome P_{450} system (see Section 2.4.2), to detoxify them. Even if such systems are available, care must be taken not to overload them. It would, therefore, seem prudent to minimize the anthropogenic production of these compounds.

8.2 Radionuclides

8.2.1 *Properties and sources*

Radionuclides are elements which spontaneously disintegrate (decay) into smaller particles and emit ionizing radiation. Certain types of this radiation, e.g. alpha particles and neutrons, have a high energy transfer per unit path length. This means that they do not travel far from the source, i.e. no more than 70 μm in soft tissue, but have pronounced effects on any atom or molecule with which they come in contact. Even at low doses the effects can be very serious, but the risk is highly localized. Beta particles (electrons) can travel farther, i.e. up to a few mm, but are still not very penetrating. Their effects at low doses can also be serious. On the other hand, high energy X- or gamma-rays are very penetrating but, because of their low energy transfer per unit path length, tend to have little effect at low doses, e.g. the 27 to 30 millirads usual in medical chest X-ray. The rad is defined as 0.01 J kg^{-1}, i.e. it is a measure of the amount of radiation energy absorbed per unit mass of material. However, one rad of alpha-radiation is about four times as damaging as one rad of X-rays. Therefore, in order to compare doses in terms of biological effects, another unit is used, the rem. One rem is biologically equivalent to one rad of X-rays. As the rad and rem are very large units relative to normally occurring radiation, most of this is measured in millirads or millirems. Recently, SI units for radiation have been introduced. In these, the gray (Gy) replaces the rad, and the sievert (Sv) replaces the rem. The relationship between these units is shown in Table 8.1.

Table 8.1   Units for measurement of radioactivity.

Physical quantity	SI unit	Non-SI unit	Relationship
Activity, i.e. nuclear transformations	becquerel (Bq) $1\ Bq = 1\ s^{-1}$	curie (Ci)	$1\ Bq = 2.7 \times 10^{-11}\ Ci$ $1\ Ci = 3.7 \times 10^{10}\ Bq$
Absorbed dose	gray (Gy) $1\ Gy = 1\ J\ kg^{-1}$	rad	$1\ Gy = 100\ rads$ $1\ rad = 0.01\ Gy$
Dose equivalent	sievert (Sv) $1\ Sv = 1\ J\ kg^{-1}$	rem	$1\ Sv = 100\ rems$ $1\ rem = 0.01\ Sv$

The disintegration rate of a radionuclide is defined in terms of its half life, which is the time taken for half the atoms in a given sample to disintegrate. Radionuclides with short half lives, i.e. up to a few days, may be very dangerous when produced, but the danger does not last. Those with very long half lives, i.e. more than 10^5 years, are usually of such low activity as to be fairly safe. Therefore, those with intermediate half lives have the greatest environmental significance.

Radionuclides occur naturally in the environment. In the UK the average person is exposed to 100 millirads of radiation from natural sources each year. Radionuclides are also produced directly or indirectly from human use of uranium. As a result of mining and processing ores to produce usable radioactive materials, large piles of waste 'uranium tailings' are left behind and radionuclides can be leached from these by rainfall, etc. Atmospheric explosions of nuclear weapons produce fallout of radionuclides over large areas of the earth's surface. Two components of this fallout have caused considerable concern, i.e. strontium-90 and caesium-137 (see Section 8.2.2). In nuclear power plants radionuclides are formed when impurities in the primary coolant water are bombarded with neutrons from the fuel elements in the core. Even if the coolant water is free of impurities radionuclides may escape into it from the steel or zirconium containers holding the radioactive fuel. Thus, the coolant water must be constantly monitored so that release of radionuclides into the environment can be minimized. With time, radionuclides produced by nuclear fission accumulate in the nuclear reactor fuel and slow or stop the fission reactions. At this point, the fuel elements must be removed and processed to extract the fission products. The disposal of these nuclear wastes presents considerable problems. Liquid and slurry wastes from reprocessing are stored in steel or concrete tanks until the radioactivity falls to acceptable levels for discharge and dispersal, often into the sea, but, as some of the nuclides are very long-lived, there could be a long-term problem caused by their accumulation, possibly far from the point of discharge. For example, caesium 137 from the discharge of Windscale is accumulating, at present at very low levels, in various flatfish in the North Sea. Solid materials, with low levels of radioactivity, may be mixed with cement or other immobilizing agents and sealed in steel drums for burial, but some radioactivity may still remain when the drums eventually corrode. High level materials present severe problems caused by the high temperatures resulting from the radioactivity, by the longevity of some of the products (half lives of thousands of years) and by their toxicity. The most promising method of disposing of them seems to be vitrification, i.e. conversion to a glass-like product, followed by burial at a great depth under the earth's surface. Very low level discharges of radionuclides may be made directly into sewers or into the atmosphere from high chimneys. A list of some important radioactive isotopes with their properties is given in Table 8.2.

Table 8.2 Half life and radiation characteristics of some important radioactive isotopes (from *The Radiochemical Manual*, 2nd edition, edited by B. J. Wilson, *The Radiochemical Centre*, Amsterdam, 1966).

Atomic number	Symbol and mass number	Half life	Radiation characteristics*
1	^3H	12.26 years	$\beta^-0.018$
6	^{14}C	5760 years	$\beta^-0.159$
15	^{32}P	14.3 days	$\beta^-1.71$
16	^{35}S	87.2 days	$\beta^-0.167$
20	^{45}Ca	165 days	$\beta^-0.254$
38	^{90}Sr	28 years	$\beta^-0.54$
53	^{131}I	8.04 days	$\beta^-0.25, 0.33, 0.61, 0.81$ $\gamma0.08, 0.18, 0.36, 0.64, 0.72$
55	^{137}Cs	30 years	$\beta^-0.51, 1.17$
94	^{239}Pu	24 400 years	$a5.096, 5.134, 5.147$

* a, alpha particles; β^-, negative beta particles; γ, gamma rays. Figures show maximum values of energies of radiation in million electron volts.

8.2.2 *Effects*

When ionizing radiation from a radionuclide hits an atom it strips off an electron and releases a large amount of energy. Thus, almost any biological molecule hit directly by ionizing radiation is functionally destroyed. This, in itself, is enough to cause considerable damage to a living cell but the damage is accentuated by the formation of highly reactive free radicals. Since water is the main component of any cell, most of these radicals are either H or OH. Radiation damage of any kind is greatest in the presence of oxygen.

Direct damage to DNA, or indirect damage from reaction with free radicals, causes chromosome breaks, cross-linking of the polynucleotide chains or modification of the constituent bases. Various repair processes can occur but there is still a high probability of mutations being produced. Most of these mutations are lethal and the rest are frequently teratogenic or carcinogenic.

Damage to proteins is associated with a general fall in enzyme activity which has been attributed mainly to the splitting of disulphide bonds. Hydrolytic enzymes may appear to become more active, but this almost certainly reflects their release from lysosomes, following damage to lysosomal membranes. Once released, they rapidly destroy the cell.

At the whole plant or animal level, the net effect of ionizing radiation is to increase mortality. This follows from harmful mutations, reduced metabolism and, in higher animals, from lowered resistance to disease following depression of the immunological system.

Strontium-90 and caesium-137, in fallout from nuclear weapons, have attracted considerable attention because of their ready accumulation by living organisms. Strontium-90 is similar to calcium and becomes localized in bones, where the emitted radiation can damage the blood cell forming bone marrow and cause anaemia or leukaemia. Caesium-137 is similar to potassium and is actively absorbed by all cells. Thus, the chances of radiation damage are greatly enhanced.

Of the high level radiation products of nuclear reactors, plutonium-239 is the one which has caused the most concern. It emits α-radiation. Like strontium-90 it accumulates in bone, with similar consequences of anaemia and leukaemia. Combustion of plutonium produces plutonium dioxide particles, any one of which, if inhaled, has a high probability of causing lung cancer.

9 Assessment of Environmental Risk

When a toxicant enters the environment many things can happen to it. It may decompose fairly rapidly to form essentially nontoxic products, any harmful effects being localized in time and space. However, the toxicant or toxic derivatives may persist. Then, the extent and degree of environmental damage will depend upon the properties of the chemicals involved and the nature of the affected ecosystem. The properties of a wide range of potential environmental contaminants have already been described. In the case of atmospheric toxicants, these properties were discussed in relation to the environment in which they occurred. In this chapter, the properties of the terrestrial and aquatic environments will be outlined, with particular reference to those factors which determine the balance between persistence and degradation of toxicants and those which influence their dispersal through the biosphere. On this basis, the problems of monitoring toxicants and of assessing environmental risk will be discussed.

9.1 Toxicants in the Soil

9.1.1 Soil chemistry and toxicant persistence

Soil characteristics affecting toxicant persistence include particle size, mineral and organic matter content, hydrogen ion concentration and microbiological activity. In general, the smaller the soil particles the longer toxic substances will persist. This is because small particles provide a large surface area for absorption of chemicals, a process which usually has a stabilizing effect. On the other hand, adsorption of toxicants reduces their availability to organisms and this may compensate for the increased persistence. For soils of a given particle size, the soil type can be shown to influence persistence, which generally decreases in the following order: organic soil, sandy loam, silty loam, and clay loam.

Probably the two most important inanimate components of soil in relation to toxicant persistence are organic matter and clay. The organic matter consists mainly of humic compounds which have a very high cation exchange capacity. Carboxyl, amino and phenolic groups provide sites for hydrogen bonding with toxicants. Sodium humate has detergent-like properties and can help to solubilize water-insoluble compounds such as DDT.

Clay is the name given to the smallest soil particles (about 0.002 mm in diameter). Clay soils are defined as those with more than 40% of their particles of this size. Clay and organic matter are often associated in soil colloids which can absorb toxicants strongly. The degree of adsorption depends upon the soil pH, moisture content, mineral ion content, temperature and any other factors which influence the physicochemical state of the colloids or toxicants. The electrical potential of the colloid surface is particularly important.

Colloid adsorptive capacity is not the only property influenced by the soil ion content. Hydrogen ion concentration (pH) controls the stability of minerals, soil ion exchange capacity, the rates of chemical reactions and microbial growth and metabolism. Cations affect soil structure by influencing flocculation and dispersal of the colloids. Some metal ions, e.g. those of iron, aluminium and magnesium, can act as catalysts, accelerating breakdown or transformation of susceptible toxicants. This is one of the many areas in which more research is needed.

9.1.2 *Metabolism of toxicants by soil organisms*

Much of the degradation and modification of chemicals in the soil is due to bacteria and fungi. These processes are facilitated by any factors which promote metabolic activity of the micro-organisms. Optimal growth temperature, pH and nutrient availability are particularly important. In some cases, oxygen may be required, while in others its absence may be necessary. Before taking any steps to promote microbial transformation of toxicants, it is important to ensure that the products of transformation are not more toxic than the original chemicals, as, for example, in the conversion of inorganic mercury to methyl mercury (see Sections 4.5 and 6.2.3).

Soil-dwelling invertebrates, such as mites and earthworms, may contribute to the chemical modification of toxicants. Free enzymes in the soil may also have a role to play. These enzymes come from dying organisms, plant roots and excreta. Since evidence suggests that less than 50% of the known decomposition of pesticides in the soil can be attributed to micro-organisms or nonenzymic chemical reactions, the contribution to this made by invertebrates and free enzymes may be considerable.

9.2 Toxicants in Natural Waters

9.2.1 Water chemistry and toxicant persistence

Most of the world is covered by seawater, with a salinity of 35 parts per thousand. Most of the remaining water is classified as fresh water, though rarely as pure water. The solvent power of water ensures that it usually contains dissolved carbon dioxide, oxygen and nitrogen, and metals such as sodium, magnesium, calcium and iron.

Water soluble toxicants are rapidly transferred to natural waters, either by leaching from the soil or by precipitation from the atmosphere. Their removal from the water then depends on their other chemical properties. Some may decompose spontaneously or volatilize. Others may form insoluble salts which precipitate, and are incorporated into sediments. Adsorption on to particulates may lead to similar consequences, or it may facilitate ingestion by filter feeders. Uptake by aquatic organisms may be followed by metabolism of toxicants to derivatives of either greater or less toxicity and their subsequent accumulation or excretion. Since many aquatic organisms have the ability to concentrate solutes, without any obvious damage to themselves, they may act as toxicant amplifiers, making the toxicants available to predators at dangerously high concentrations. The predators may accumulate even higher tissue concentrations then the prey because they consume large numbers of them. This process of toxicant amplification can continue to the top of any food chain, provided that the toxicant is chemically stable and has some property, e.g. lipid solubility which inhibits its excretion. Food chain amplification of this kind is observed with sacitoxin (see Section 4.4.1), chlorinated hydrocarbons (see Sections 5.2.1, 5.4 and 5.5), and methyl mercury (see Section 6.2).

Persistence of toxicants in natural waters reflects their chemical properties. Hydrophobic substances, with appreciable lipid solutility, tend to be the most resistant to spontaneous chemical modification or metabolic change and have the greatest tendency to accumulate in organisms. However, they also tend to accumulate at the water surface. This facilitates their loss by volatilization and exposes them to ultraviolet radiation from the sun which may promote their decomposition.

9.2.2. Metabolism of toxicants by aquatic organisms

The role of micro-organisms in the aquatic environment in degrading toxicants is at least as important as in the terrestrial environment. Not only are they involved in the metabolism of toxic substances but some produce toxins of their own*(see Chapter 4). In addition they may cause the death of other organisms as a result of localized oxygen depletion. This accompanies bacterial decomposition of the oxygen demanding wastes which occur in sewage, or effluent from food processing plants, paper mills, tanneries, slaughter houses and intensive animal production units. The effects of oxygen depletion may be aggravated by toxicants, e.g. amines and hydrogen sulphide (see Section 4.5) which are the products of anaerobic breakdown of these wastes. Similar effects of oxygen depletion and toxicant production accompany the decomposition of algal blooms following excessive eutrophication (see Section 9.2.3).

Metabolism of toxicants by aquatic fauna follows the general principles outlined in Chapter 2, but much remains to be learned, both about this and about the metabolism of aquatic plants. Since the micro-algae, or phytoplankton, contribute about one third of the total world primary production, harmful effects on them could be particularly serious. However, so far there have been few reports of such damage and any that has occurred seems to have been highly localized.

9.2.3   Eutrophication

Eutrophication simply means enrichment with nutrients, and its occurrence in natural waters is therefore essential for the maintenance of life. Problems arise when eutrophication is excessive. This state is sometimes referred to as cultural eutrophication. Discharge of nutrients (e.g. nitrate, phosphate) or limiting organic compounds, (e.g. vitamins) in sewage, industrial effluents or runoff from fertilizer-rich agricultural land, can result in such an excess of nutrients in inland lakes and ponds that certain algae flourish abnormally. These algae include the blue-green species *Aphanizomenon flos-aquae*, *Anabaena spiroides*, and *Oscillatoria rubescens*. Initially, algal photosynthesis oxygenates the water but, as nutrients run out and the algae die, their decomposition is accompanied by oxygen depletion (see Section 9.2.2) which leads to almost complete elimination of the normal fauna. The production of toxins by some of the blue-green algae (see Section 4.4) serves to accelerate the process. Although generally characteristic of inland waters, it may be that excessive eutrophication is in its early stages in some semi-enclosed marine systems, such as the Baltic, the North Sea and the Mediterranean.

9.2.4   Salinity

Salinity may be an important toxic factor in the aquatic environment. Many fresh water organisms are killed if their habitat becomes brackish. The converse is also true, but of less general importance since the tendency of water is always to acquire salts rather than to lose them. The main effects of increasing salinity are due to the associated increase is osmotic pressure of the water. Hence, they are very similar to the effects of dehydration.

Fig. 9.1 Salination of irrigated desert in the Punjab. The salt comes originally from rain which evaporates from the land before the salt it carries can be washed away. These salts accumulate over the years and are brought to the surface as the water table rises following irrigation. Only much improved drainage with deep channels or the growth of salt absorbing vegetation will make this land fit for cultivation again (from the Diana Wyllie slide set 'Air Pollution).

Salinity may also be a problem in the terrestrial environment as it affects irrigation of agricultural land (Fig. 9.1). Irrigation water leaches salts from the soil and, if this water is recycled, the salt concentration may become high enough to inhibit crop growth. A recent estimate suggested that 25% of the irrigated land in the USA is now affected to some extent by this phenomenon.

9.3 Effects of Temperature and Weather Conditions

9.3.1 Temperature
Raising temperature increases the rate of volatilization of chemicals, their water solubility and the rates of reactions in which

they are involved. Within limits, it also increases the rate of uptake of toxicants by living organisms and promotes metabolism. These limits vary from one organism to another, but usually approximate to the temperatures for optimum growth. Thus, psychrophiles respond best at temperatures below 20°C, mesophiles between 20 and 45°C and thermophiles between 45 and 65°C. In plants, translocation increases with temperature, as does loss to the atmosphere of volatile compounds in transpiration or through the cuticle. In animals, increased superficial circulation may facilitate loss of toxicants through the skin, lungs or gills, though reduced kidney excretion may counteract this.

Increased environmental temperatures are usually the result of increased radiation in the form of sunlight. The ultraviolet content of this radiation promotes the breakdown of exposed toxicants, but may also have mutagenic effects by damaging DNA in exposed cells. These mutagenic effects may be beneficial, if they destroy harmful viruses or bacteria, or they may be detrimental, if they cause tumours. Ultraviolet radiation also facilitates vitamin D synthesis in exposed skin. Vitamin D is essential for adequate absorption of calcium from the gut in developing animals and, hence, for proper bone formation.

Another cause of increased environmental temperatures in aqueous habitats is the use of water as a coolant in industrial processes (Fig. 9.2) Used coolant water may have a temperature 12°C higher than the river, stream or bay to which it is returned. One of the most obvious consequences of the discharge of warm water is a decrease in the amount of dissolved oxygen near the discharge point. This is because hot water cannot dissolve as much oxygen as cold water and, being lighter than cold water, forms a blanketing layer on the surface. Increased water temperature also increases the respiration rate of marine organisms. Thus, they require more oxygen in an environment which contains less than before. As a result, they may suffocate. Hot summer days may have the same effect.

Discharge of warm water from cooling systems may have more subtle effects by giving false temperature cues to aquatic organisms, hence causing inappropriate behaviour for the time of year, e.g. migration and spawning may be adversely affected. Many molluscs spawn within a few hours after their environment reaches a critical temperature. Normally this will coincide with suitable conditions for larval development. If it occurs at the wrong time, fertilization may be inadequate or larval maturation may fail. Similar considerations apply to most other organisms.

Even organisms which are normally tolerant of elevated temperatures may be adversely affected by warm water discharges.

Fig. 9.2 A steaming river—the River Severn at Ironbridge, England. When this photograph was taken, the power station had no cooling towers and the cooling water went straight into the river producing this effect (from the Diana Wyllie slide set 'Air Pollution').

In this case, the ill effects may be traced to over-rapid changes in temperature which do not allow time for acclimatization. Other ill effects may result from additives, such as chlorine or, in the case of nuclear power plants, from the presence of radionuclides (see Section 8.2). Additives are put in to prevent bacterial growth and sometimes to prevent corrosion in the pipes.

Minimizing the effects of warm water discharges can be accomplished in one of two ways. Either cooling towers may be used to remove excess heat before discharge, or the warm water may be temporarily stored in a cooling pond. The former process is more expensive to install, but it does present the possibility of using the heat for other purposes instead of releasing it wastefully into the environment.

9.3.2   *Weather conditions*

The effects of sunlight have been largely described in Section 9.3.1 In this section, attention will be concentrated on precipitation, humidity and air movement.

Precipitation, in the form of rain, hail or snow, may carry down toxic substances from the atmosphere, as described in Chapter 7, and deposit them on exposed surfaces. Absorption of such toxicants by plants will be favoured by the wet conditions, but other toxicants may be washed off. Within limits, increase in soil water by precipitation increases inherent biological activity and, therefore, promotes the metabolism of toxic substances. Flooding kills many organisms, but facilitates anaerobic bacterial transformations. It also releases many chemicals from binding to soil colloids, resulting either in their loss by runoff to surface water or their transfer to plants or animals.

Precipitation is associated with increased air humidity, which reduces volatilization of chemicals from the soil and from plants. Guttation in plants, the exudation of water droplets from the leaves, is restricted under these conditions. If the humid atmosphere is in contact with dry soil, water soluble compounds will diffuse with absorbed moisture towards the deeper soil layers. When the soil is moist and the air is dry, this process will be reversed.

Any increase in air movement increases the loss of volatile chemicals from exposed surfaces. If the air movement is sufficient, it may remove surface deposits of particulates (see Section 7.4) or spray droplets. Where leaves are affected, closure of stomata may decrease uptake of toxicants.

9.4 Distribution of Toxicants in the Biosphere

No matter how toxic a substance may be, its presence in the environment is of concern only if it affects, or is likely to affect, living organisms. In most cases, this depends upon its uptake by certain of these organisms, and subsequent distribution through the biosphere by way of predator-prey relationships in food webs. Details of the physiological processes involved have been described in Chapters 2 and 3, and the process of food chain amplification is described in Section 9.2.1. A number of attempts have been made to incorporate this knowledge into a predictive model in order to assess the likely risk from new toxicants in the environment. The simplest approach is to consider toxicant distribution as a series of partitions between environment and organism, organism and predator, and so on. Given a constant level of toxicant in the environment, a stable ecosystem and sufficient time, a steady state distribution should be attained. Under these circumstances, each organism in the contaminated environment will contain the toxicant in proportion to a characteristic partition coefficient between itself and the environment, no matter how many intermediaries there may be. Furthermore, the localized toxicant concentration in specific parts of the organism will be nearly constant, though the total will alter with the size of the part. Again partition coefficients may be determined to define mathematically the distribution between the tissue or cell compartments. Since metabolities partition independently of the parent substance, metabolism of a toxicant may serve only to distribute derivatives throughout the affected organism or, after excretion, throughout the environment. Metabolism cannot reduce levels of toxicant in an organism exposed to a constant external concentration.

The partition approach has been applied successfully to mathematical analysis of closed laboratory systems, but extrapolation to the natural environment has proved difficult for the following reasons. Levels of toxicants do not normally remain constant. Living organisms go through developmental stages of varying susceptility to different chemicals and with varying capacities for their absorption. Short lived organisms may not survive long enough to equilibrate with the environment. Animals may migrate from contaminated environments to clean environments and back again. Plant seeds may be dispersed or lie dormant in the soil. Mutations may occur, and toxicant resistant organisms appear. Because of these difficulties, it seems likely that mathematical models of toxicant distribution in the biosphere will not be of value in environmental planning and management for some time to come. Their chief value at present is probably in generating quantitative hypotheses that can be tested by experiment or observation.

9.5 Monitoring Environment Toxicants

9.5.1 *Chemical analysis*

Until such time as accurate predictions of the fate of toxicants in the environment can be made, reliance must be placed on well-designed monitoring schemes to provide the necessary information for appropriate steps to be taken in the prevention of environmental damage. Monitoring requires good methods of chemical analysis. Often these are available, but it may not be clear what chemicals the analyst should try to detect. In addition to the suspected toxicant, there will be its metabolites of varying toxicity. There may also be synergists and antagonists (see Section 2.5). Having decided what to look for, the analyst must know what levels are to be regarded as dangerous and adjust the sensitivity of his methods appropriately. Unnecessary analysis of trace quantities of chemicals can be an expensive waste of manpower and materials. However, establishing safe limits for toxicant exposure is not easy. The practical difficulties were discussed in Chapter 1. The conceptual difficulties may be even greater. The natural tendency is to define limits primarily in relation to direct risk to human beings. This may include risk to agriculturally important plants and animals, economically important fish, domestic pets and garden plants. However, there are many other organisms, each playing a part in the total world ecosystem, and our knowledge is as yet insufficient to predict the consequences of partial or complete elimination of many of these species. Thus, while practical necessity makes it essential to define safe limits in relation to those organisms of which most is known, such limits must be open to modification as further knowledge is gained. In cases where no absolutely safe limits can be defined, it may be necessary to establish levels below which risk, though present, can be regarded as acceptably small. This raises the question of what constitutes an acceptable risk. In some cases, e.g. the carcinogens in cigarette smoke, real or imagined benefits associated with the toxicants may make quite a high risk acceptable to *Homo sapiens*.

Even if the chemical analyst knows what to look for and what concentrations are significant, there remain problems of sampling and extraction. Sampling may be random throughout the environment, or selective or a mixture of both. The sampling programme must be designed to be statistically sound where comparisons are being made. Sampling methods must be chosen which are effective in preventing loss of material between collection and analysis, and which avoid contamination of samples with extraneous material. Time between sampling and analysis should be kept to a minimum, or well-established methods of sample and toxicant preservation used to prevent post-sampling changes. Having obtained adequate samples, the analyst must extract the toxicants present. Whatever extraction method is used, some toxicants will be left behind. The question arises as to whether the unextracted toxicants can be safely ignored. The answer to this depends on the relationship between extractability and biological availability. Normally the two will be closely related, but this may not always be the case. In dealing with higher organisms, it may often be better practice to extract selected parts rather than the whole organism. This is because a high localized concentration in one tissue may be masked when results are expressed on a whole organism basis. How to express results is always a difficult decision. Not only can the amount of toxicant be expressed per organism, or per tissue, but it can be expressed per cell, per gram wet weight, per gram dry weight, per cm^3 body, tissue or cell water, etc. Similar problems arise with environmental material.

Perhaps the most important question for the chemical analyst is what is the minimum quantity of toxicant that can be detected reliably, since this is a measure of the technical limitations that are imposed on environmental monitoring. Awareness of these limitations is essential, because the analyst can never prove that toxicants are absent from the environment but can only demonstrate their presence at levels which respond to his techniques.

9.5.2 *Biological analysis*

If assessing hazardous levels of chemicals in the environment is difficult, assessing environmental deterioration is even more so. The first problem is in establishing what constitutes normality in the area of interest. This requires what has been called a base line survey. Ideally, it should involve a control area identical to the environment at risk, but finding such an area may well be impossible. Thus, one has to settle for a similar area, with the error that this entails. Thereafter, a survey is made of the species inhabiting the area and of their relative abundance. It is important that the methods used for assessment are objective and of similar efficiencies in both locations. Transects are often used for studying the vegetation and less mobile animals. They are particularly useful if they can be aligned to a known gradient of toxicant. Otherwise, random quadrats are preferable. Rare species, because of their rarity, may be overlooked in such surveys. This is unsatisfactory because they may be particularly susceptible to toxicants. However, from the information gained on species and their relative abundance, a diversity index may be computed (e.g. Krebs, 1978). If the index is low, this is often regarded as a sign of environmental stress but it may reflect other factors, including the homogeneity or heterogeneity of the habitat, its variability with time and rates of predation. Thus, spatial and temporal fluctuations of each species in the control area and the area at risk must be determined to establish their relationship to normal environmental factors and their biological interactions. In this way, a total picture of both areas will be obtained and differences may be observed. The significance of these differences must now be determined and the following questions answered: how many species have to be absent from the contaminated environment relative to the control, and for how long, before damage can be established with certainty; can damage be inferred where species have not disappeared from the contaminated area although their numbers have been significantly changed? Answers to these questions cannot be easily obtained, and those that have been put forward must frequently be viewed with scepticism in our present state of ignorance.

9.5.3 Critical path analysis

Given that effective monitoring is possible, much time and effort can be saved if a critical path through the environment can be identified. Such a path may be defined as the one from which the greatest harm may be anticipated. It is likely that, somewhere on this path, one can identify a critical group (see Section 1.3.6), protection of which will ensure that no other organisms are subject to risk. For example, certain organisms may accumulate heavy metals without apparent harm to themselves, but putting their predators, including human beings, at risk. If these organisms are monitored and their environment conserved so as to keep their heavy metal content below the danger level for their predators, no harm should ensue to the latter or to other organisms that are less at risk. Thus, the environment may be protected without having to monitor every aspect.

9.6 Conclusion

The aim of the environmental toxicologist is to provide information about the risks posed by the presence of natural or anthropogenic toxicants in the environment. This information is subject to two important limitations. It is impossible to prove the absence of a toxicant, and it is impossible to prove that low concentrations of toxicants are without effect on living organisms. Thus, any decision on what constitutes a safe level of a toxicant is to some extent a guess. Nevertheless, if such a decision is based on reliable information, it will at least be a good guess. Obtaining reliable information is not easy. There are many problems, which this book has sought to define. There are few solutions, but no problem can be solved until it has been clearly expressed, and people are aware that it exists.

Appendix 1

The Relationship of Toxicity to Thermodynamic Activity

Ferguson in 1939 suggested that the relationship of toxicity to thermodynamic activity might be more consistent than its relationship to concentration. Thermodynamic activity is defined as c/c_0 where c is the concentration of a solute whose solubility in the solvent being used is c_0. The physical effect on a system of any toxicant at a given activity is exactly the same as that of any other toxicant at the same activity, though the relationship to concentration will vary. Thus, if toxicity bears a constant relationship to activity for a range of chemically distinct toxicants, it is likely that the fundamental effect is physical.

Besides being useful in distinguishing between physical and chemical toxicity, knowledge of the toxicity/activity relationship is invaluable in dealing with a multiphase system. In such a system, at equilibrium, the activity of any solute will be the same in all phases. Thus, it need only be determined in one. Since any living cell is a multiphase system, the importance of this relationship is obvious, though its application is restricted to those toxicants which partition passively between the environment, the cell and its organelles.

Appendix 2

Environmental Indices

An index is a number expressing the relationship of a variable quantity to a fixed standard. Such a number can be used to simplify the presentation of data so that detailed knowledge is not needed for their interpretation. Thus, an index provides a means of making specialized knowledge available to the nonspecialist. Where matters are as complex as in the environment, this is particularly important.

Numerous indices have been suggested to define the state of the environment but, to illustrate the principles involved, only one will be described here. In 1975, the US Council on Environmental Quality (CEQ) set up the Federal Interagency Task Force on Air Quality Indicators to consider the air quality indices then in use in the USA, and to recommend a standard index to replace them. The result was the Pollutant Standards Index (PSI or φ) shown in Table A.1. The following criteria were applied in compiling this index:

(1) It must be easily understood by the public
(2) It must include the currently accepted pollutants, or toxicants, and be able to include pollutants, or toxicants, which may be discovered in the future
(3) It must relate to the US National Ambient Air Quality Standards (NAAQS)
(4) It must relate to Federal Episode Prevention Criteria
(5) It must be easily calculated
(6) It must be consistent with perceived air pollution levels
(7) It must be able to be forecast a day in advance if required.

The index, as formulated, includes five toxicants, i.e. total suspended particulates (TSP), sulphur dioxide, carbon monoxide, ozone (or oxidants) and nitrogen dioxide. In the USA, National Ambient Air Quality Standards, Federal Episode Criteria, and Significant Harm Levels exist for all five.

The PSI is set arbitrarily a 100 for each National Ambient Air Quality Standard level and at 500 for each Significant Harm Level. Another gradation is added at 50% of the primary National Ambient Air Quality Standard level for each toxicant, excepting sulphur dioxide and total suspended particulates, where secondary National Ambient Air Quality Standards are used. This provides an intermediate value between marginal and good air quality. Five descriptive categories are used in the PSI to clarify the significance of the index figures when published in the press, along with an outline of general health effects and cautionary statements, all of which are shown in Table A.1.

Environmental indices, such as the PSI, are clearly of great potential value in alerting the public to dangerous trends, but they cannot be any better than the data on which they are based. Such data are subject to many provisos, no matter how carefully they are compiled. One objection to indices is that these provisos tend to be forgotten once the index has been compiled. Similarly, establishing an index implies the existence of a suitable standard for reference, and it has been pointed out in Chapter 9 that such standards are set by informed guesswork. However, once an arbitrary standard is enshrined in an index, especially a national or international one, its doubtful origins, like those of the data to which it will be compared, may well be ignored. This is not an argument against indices, however, but rather a case for ensuring that, once established, they are periodically revised in the light of current knowledge.

Table A.1 Comparison of φ values with pollutant concentrations, descriptor words, general health effects and cautionary statements (from *The Sixth Annual Report of the Council on Environmental Quality—1975*, Council on Environmental Quality (US), Washington DC, 1975).

Index value	Air quality level	Pollutant levels					Health effect descriptor	General health effects	Cautionary statements
		TSP (24 hour) $\mu g\ m^{-3}$	SO_2 (24 hour) $\mu g\ m^{-3}$	CO (8 hour) $mg\ m^{-3}$	O_3 (1 hour) $\mu g\ m^{-3}$	NO_2 (1 hour) $\mu g\ m^{-3}$			
500	Significant harm	1000	2620	57.5	1200	3750		Premature death of ill and elderly. Healthy people will experience adverse symptoms that affect their normal activity	All persons should remain indoors, keeping windows and doors closed. All persons should minimize physical exertion and avoid traffic
400	Emergency	875	2100	46.0	1000	3000	Hazardous	Premature onset of certain diseases in addition to significant aggravation of symptoms and decreased exercise tolerance in healthy persons	Elderly and persons with existing diseases should stay indoors and avoid physical exertion. General population should avoid outdoor activity

300	Warning	625	1600	34.0	800	2260	Very unhealthful	Significant aggravation of symptoms and decreased exercise tolerance in persons with heart or lung disease, with widespread symptoms in the healthy population	Elderly and persons with existing heart or lung disease should stay indoors and reduce physical activity
200	Alert	375	800	17.0	400c	1130	Unhealthful	Mild aggravation of symptoms in susceptible persons with irritation. Symptoms in the healthy population	Persons with existing heart or respiratory ailments should reduce physical exertion and outdoor activity
100	Moderate	260	365	10.0	160	a	Moderate		
50	50% of NAAQS	75b	80b	5.0	80	a	Good		
0	0	0	0	0	0	a			

a No index values reported at concentration levels below those specified by alert level criteria
b Annual primary NAAQS.
c 400 μg m⁻³ was used instead of the ozone alert level of 200 μg m⁻³.

Appendix 3

Environmental Impact Statements

Environmental indices usually describe the current situation over a fairly large area. Environmental impact statements on the other hand, are required to predict the effects of specific projects on defined, and often small, areas, such as a bay into which effluent may be discharged from a power station or a new factory. Because they try to evaluate every aspect of the environment, they tend to be less precise and more subjective than indices. At their simplest, impact statements may be compiled by putting together all

	CHARACTERISTICS						
ACTIONS	Lead in air	Wild mammals	Outdoor recreation	Rare plants	Turbidity in water	Radioactivity	Noise
Canal building			3	1	6		
Automobile traffic	10		2				8
Swamp drainage		7	5	3	1		
Can recycling			2				
Pulp and paper mills			7	2	8		5
Park construction		10	10	7	2		

Fig. A.1 The matrix approach to environmental impact statements. One corner of a matrix is shown. Each box represents an estimate of degree of change in a given environmental characteristic resulting from a specific action. The higher the value in the box the greater the effect. No attempt has been made to indicate in this matrix whether the effects are beneficial or harmful. For clarity harmful effects may be designated by a negative sign (after H. Inhaber, from Environmental Indices, Wiley Interscience, New York; Figure 20).

available information, and subjecting it to arbitrary value judgements. This approach may be rationalized by constructing a matrix, as shown in Fig. A.1. Each box in the matrix represents the effect of an action upon an environmental characteristic, expressed as a value between 0 and 10 where 10 represents the maximum effect, whether harmful or beneficial. Where the value is 0, the box is left empty. harmful effects may be indicated with a negative sign. Such a matrix helps to clarify an impact statement, but all of the numbers arrived at still reflect the opinion of the compiler and so are highly subjective and open to debate.

The least subjective type of impact statement is that used by Battelle, which involves a large component of objective measurement. Measurable parameters of the environment are considered and a graph, or value function, is drawn for each (Fig. A. 2). The graph shows environmental quality ranging from 0, which is very poor, to 1, which is excellent, plotted against a scale

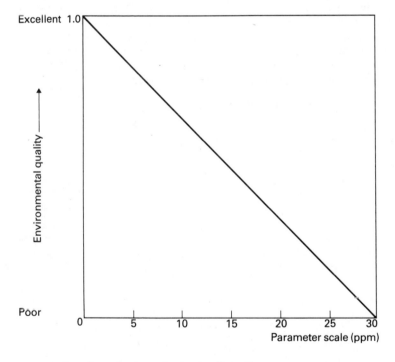

Fig. A.2 Battelle environmental value function. The parameter scale may be in any appropriate units. Given the concentration of the parameter, the 'environmental quality' can be read from the graph. The relationship between environmental quality and parameter concentration need not be linear.

representing the parameter in suitable chemical or physical units. Clearly, the assessment of environmental quality in this context requires an arbitrary value judgement, i.e. informed guess. A further value judgement, of parameter importance, is required to assess the environmental impact of a parameter using the following formula:

Environmental impact = parameter importance × environmental quality

It will be seen that this approach differs from the matrix approach, in that a Batelle environmental impact figure of zero would represent the worst possible situation, whereas a matrix figure of zero indicates no effect. In the Battelle scheme, environmental impact of a parameter is calculated for the prevailing circumstances before a project is undertaken, and estimated for those resulting from the project. If the impact is expected to decrease, the project is regarded as being harmful.

While the quantitative aspects of the Battelle approach are to be applauded, the use of the term environmental impact with the implication that any impact must have a positive effect on environmental quality can only make for confusion. Common sense dictates that an impact can be beneficial or harmful. The Battelle equation might, therefore, be better expressed as:

Beneficial impact = parameter importance × environmental quality

Appendix 4

The Chemical Nomenclature of Pesticides and Herbicides mentioned in the Text by Trivial Names†

Trivial name	Chemical name
Aldrin	1,2,3,4,10,10-Hexachloro-1,4,4α,5,8,8α-hexahydro-*endo-exo*-1,4,5,8-dimethanonaphthalene
Amitrole	3-Amino-*s*-triazole
Antu	1-(1-Naphthyl)-2-thiourea
Atrazine	2-Chloro-4-(ethylamino)-6-(isopropylamino)-*s*-triazine
Baygon	O-isopropoxyphenyl methylcarbamate
BHC	γ-1,2,3,4,5,6-Hexachlorocyclohexane
Bordeaux mixture	Copper sulphate and calcium hydroxide
Bromacil	5-Bromo-3-*sec*-butyl-6-methyluracil
Bromoxynil	3,5-Dibromo-4-hydroxybenzonitrile
Captan	N-(trichloromethylthio)-4-cyclohexene-1,2-dicarboximide
Carbaryl	1-Naphthyl N-methylcarbamate
Chloranil	2,3,5,6-Tetrachloro-*p*-benzoquinone
Chloropicrin	Trichloronitromethane
Chloropropham	Isopropyl-*m*-chlorocarbanilate
Compound 1080	Fluoroacetic acid, sodium salt
Cycloheximide	3-[2-(3,5-dimethyl-2-oxocyclohexyl)-2-hydroxyethyl] glutarimide
D-Cycloserine	D-4-amino-3-isoxazolidinone
Cyprex	*n*-dodecylguanidine acetate
2,4-D	2,4-Dichlorophenoxyacetic acid
DDD	1,1-Dichloro-2,2-bis (*p* chlorophenyl) ethane
DDE	1,1-Dichloro-2,2-bis (*p*-chlorophenyl) ethylene
DDT	1,1,1-Trichloro-2,2-bis (*p*-chlorophenyl) ethane

† Synonymous trivial names are listed alphabetically in the index with a cross reference to the trivial name used in the text.

Trivial name	*Chemical name*
Diazinon	O,O-diethyl O-(2-isopropyl-6-methyl-4-pyrimidinyl) phosphorothioate
Dichlone	2,3-Dichloro-1,4-naphthoquinone
Dichloromate	3,4-Dichlorobenzylmethylcarbamate
Dieldrin	*Endo, exo*-1,2,3,4,10,10-hexachloro-6,7-epoxy-1,4,4α,5,6,7,8,8α-octahydro-1,4:5,8-dimethanonaphthalene
Diethylcarbamazine	N,N-diethyl-4-methyl-1-piperazinecarboxamide
Dinoseb	2-*sec*-butyl-4,6-dinitrophenol
Dioxin	2,3,6,7-Tetrachlorodibenzo-*p*-dioxin
Diphacin	10-(N,N-Diethylglycyl)phenothiazine
Diquat	6,7-Dihydropyrido[1,2-α:2′,1′-*c*]-pyrazidinium dibromide
Diuron	3-(3,4-Dichlorophenyl)-1,1-dimethylurea
DNOC	4,6-Dinitro-*o*-cresol
Dodine	See cyprex
Ferbam	*Tris* (dimethyldithiocarbamato) iron
Fumarin	3-(α-Acetonylfurfuryl)-4-hydroxycoumarin
Glyphosate	N-phosphonomethylglycine
Haloxydine	3,5-Dichloro-2,6-difluoro-4-haloxypyridine
HCB	Hexachlorobenzene
Heptachlor	1,4,5,6,7,8,8-Heptachloro-3α,4,7,7α-tetrahydro-4,7-methanoindene
Ioxynil	4-Hydroxy-3,5-diiodobenzonitrile
Isocil	5-Bromo-3-isopropyl-6-methyluracil
Karathane	2-(1-methylheptyl)-4,6-dinitrophenyl crotonate
Kepone	Decachlorooctahydro-1,3,4-metheno-2H-cyclobuta[cd]pentalen-2-one
Lindane	See BHC
Malathion	S-(1,2-dicarbethoxyethyl) O,O-dimethyldithiophosphate
Maneb	[Ethylenebis (dithiocarbamato)] manganese
MCPA	4-Chloro-2-methylphenoxy acetic acid
Methyoxychlor	1,1,1-Trichloro-2,2-bis (*p*-methoxyphenyl) ethane
Mirex	1,1a,2,2,3,3a,4,5,5,5a,5b,6-Dodecachlorooctahydro-1,3,4-metheno-1H-cyclobuta[cd]pentalene
Monuron	3-(4-Chlorophenyl)-1,1-dimethylurea
Mylone	3,5-Dimethyl-1,3,5,2H-tetrahydrothiadiazine-2-thione

Trivial name	*Chemical name*
Nabam	Ethylenebis[dithiocarbamic acid] disodium salt
Norbormide	5-(α-Hydroxy-α-2-pyridylbenzyl)-7-(α-2-pyridylbenzylidene)-5-norbornene-2,3-dicarboximide
Paraquat	1,1'-Dimethyl-4,4'-dipyridylium dichloride
Parathion	O,O-diethyl O-*p*-nitrophenyl phosphorothioate
PBB	Polybrominated biphenyl
PCB	Polychlorinated biphenyl
PCP	Pentachlorophenol
Pival	2-Pivaloyl-1,3-indandione
PMP	2-Isovaleryl-1,3-indandione
Prolin	Warfarin+Sulfaquinoxaline (see separate entries)
Propanil	3',4'-Dichloropropionanilide
Propham	Isopropyl carbanilate
Pyrichlor	2,3,5-Trichloro-4-pyridinol
Sandoz 6706	4-Chloro-5-(dimethylamino)-2-(α,α,α-trifluoro-*m*-tolyl-3(2H)pyridazinone
Sevin	1-Naphthyl N-methylcarbamate
Simazine	2-Chloro-4,6-bis(ethylamino)-S-triazine
SKF525A	β-Diethylaminoethyl-diphenyl-propyl acetate
Sulfaquinoxaline	N'-2-quinoxalyl-sulfanilamide
2,4,5-T	2,4,5-Trichlorophenoxyacetic acid
TCDD	See Dioxin
Telone [Dow]†	1,3-Dichloropropene
Temik	2-Methyl-2-(methylthio) propionaldehyde O-(methylcarbamoyl) oxime
Thiabendazole	2-(4-Thiazolyl) benzimidazole
Thiram	Bis(dimethylthiocarbamoyl)disulphide
4,5',8-Trimethylpsoralen	6-Hydroxy-β,2,7-trimethyl-5-benzofuranacrylic acid δ lactone
VC-13 Nemacide	O-2,4-dichlorophenyl O,O-diethylphosphorothioate
Warfarin	3-(α-Acetonylbenzyl)-4-hydroxycoumarin
Zearalenone	6-(10-Hydroxy-6-oxo-*trans*-1-undecenyl)-β-resorcylic acid lactone
Zectran	4-Dimethylamino-3,5-xylyl methylcarbamate
Zineb	[Ethylenebis (dithiocarbamato)] zinc
Ziram	Bis (dimethyldithiocarbamato) zinc

† Not the same as Telon [Sandoz].

Bibliography

References are listed by chapter and in alphabetical order. A supplementary section giving a guide to information sources is listed at the end.

Chapter 1 Assessment of Toxicity

ANDERSON, D., (1978). An appraisal of the current state of mutagenicity testing. *Journal of the Society of Cosmetics Chemistry*, **29**, 207–223.

ARIENS, E. J., SIMONIS, A. M. and OFFERMEIER, J., (1976). *Introduction to General Toxicology*, Academic Press, New York.

AUERBACH, C., (1976). *Mutation Research—Problems, Results and Perspectives*, Chapman and Hall, London.

BOYLAND, E. and GOULDING, R. (ed.), (1968). *Modern Trends in Toxicology*, Vol 1, Butterworths, London.

BUTLER, G. C., (1978). *Principles of Ecotoxicology—SCOPE 12*, John Wiley and Sons, Chichester.

CAIRNS, J., DICKSON, K. L. and MAKI, A. W., (1978). *Estimating the Hazard of Chemical Substances to Aquatic Life*, American Society for Testing and Materials, Heyden, London.

CASARETT, L. J. and DOULL, J. (1975). *Toxicology—The Basic Science of Poisons*, MacMillan, New York.

DE BRUIN, A. (1977). *Biochemical Toxicology of Environmental Agents*, Elsevier, Amsterdam.

EISLER, R. (1972). Pesticide induced stress profiles. In: *Marine Pollution and Sea Life*, Ed. M. Ruivo, Fishing News (Books), West Byfleet, for FAO; 318–322.

FINNEY, D. J. (1971). *Probit Analysis*, 3rd edn, Cambridge University Press, Cambridge.

HOLLAENDER, A. (ed.) (1971–78). *Chemical Mutagens, Principles and Methods for their Detection*, in 5 volumes, Plenum, New York.

HUNTER, W. J. and SMEETS, J. G. P. M. (1977). *The Evaluation of Toxicological Data for the Protection of Public Health*, Pergamon, Oxford.

KILBEY, B. J. *et al.* (eds.) (1977). *Handbook of Mutagenicity Testing Procedures*, Elsevier/North Holland Biomedical Press, Amsterdam.

KLEKOWSKI, E. J. (jr) and BERGER, B. B. (1976). Chromosome mutations in a fern population: a bio-assay for mutagens in aquatic environments. *American Journal of Botany*, **63**, 239–246.

KRAYBILL, H. F. and MEHLMAN, M. A. (1977). *Environmental Cancer*, Advances in Modern Toxicology, Vol 3, John Wiley and Sons, New York.

LEE, S. D. (1979). *Assessment of Biological Effects of Environmental Pollutants*, Ann Arbor Science, Ann Arbor.

MOUNT, D. I. and STEPHAN, D. E. (1967). A method for establishing acceptable toxicant levels for fish—malathion and the butoxyethanol ester of 2,4-D. *Transactions of the American Fisheries Society*, **96**, 185–193.

NORDBERG, C. F. (ed.) (1976). *Effects and Dose—Response Relationships of Toxic Metals*, Elsevier, Amsterdam.

PAGET, G. F. (ed.) (1970). *Methods in Toxicology*, Blackwell, Oxford.

WILSON, K. W. (1975). The laboratory estimation of the biological effects of organic pollutants. *Proceedings of the Royal Society*, B, **189**, 459–477.

Chapter 2   Metabolism of Toxic Substances by Animals

BEND, J. R. and JAMES, M. O. (1978). Xenobiotic metabolism in marine and freshwater species. In: *Biochemical and Biophysical Perspectives in Marine Biology*, Ed. D. C. Malins and J. R. Sargent, Vol 4, Academic Press, London; 125–187.

BROWN, G. W. (jr) (1976). Biochemical aspects of detoxification in the marine environment. In: *Biochemical and Biophysical Perspectives in Marine Biology*, Ed. D. C. Malins and J. R. Sargent, Vol 3, Academic Press, London; 319–406.

CRAMPTON, R. F. and CHARLESWORTH, F. A. (1975). Occurrence of natural toxins in food. *British Medical Bulletin*, **31**, 209–213.

DE BRUIN, A. (1977). *Biochemical Toxicology of Environmental Agents*, Elsevier, Amsterdam.

PARKE, D. V. (1968). *The Biochemistry of Foreign Compounds*, Pergamon, Oxford.

SCHELINE, R. R. (1978). *Mammalian Metabolism of Plant Xenobiotics*, Academic Press, London.

WALDBOTT, G. L. (1973). *Health Effects of Environmental Pollutants*, Mosby, St Louis.

WALKER, C. H. (1975). Variations in intake and elimination of pollutants. In: *Organochlorine Insecticides—Persistent Organic Pollutants*, Ed. F. Moriarty, Academic Press, London.

Chapter 3   Metabolism of Toxic Substances by Plants

CORBETT, J. R. (1974). *Biochemical Mode of Action of Pesticides*, Academic Press, New York.

DODGE, A. D. (1975). Some mechanisms of herbicide action. *Science Progress*, Oxford, **62**, 447–466.

EDWARDS, C. A. (1976). Factors that affect the persistence of pesticides in plants and soils. In: *Pesticide Chemistry—3*, Ed. P. Varo, Butterworths, London; 39–56.

FORSYTH, A. A. (1968). *British Poisonous Plants*, HMSO, London.

HARBORNE, J. B. (1977). *Introduction to Ecological Biochemistry*, Academic Press, London.

HAWKSWORTH, D. L. and ROSE, F. (1976). *Lichens as Pollution Monitors*, Edward Arnold, London.

KEELER, R. F., VAN KEMPEN, K. R. and JAMES, L. F. (1978). *Effects of Poisonous Plants on Livestock*, Academic Press, New York.

LIENER, I. E. (ed.) (1969). *Toxic Constituents of Plant Foodstuffs*, Academic Press, New York.

MUDD, J. B. and KOZLOWSKI, T. T. (eds.) (1975). *Response of Plants to Air Pollution*, Academic Press, New York.

STROBEL, G. A. (1976). Phytotoxins as tools in studying plant disease resistance. *Trends in Biochemical Sciences*, **1**, 247–250.

Chapter 4   Toxic Substances Released into the Environment by Micro-organisms

ALEXANDER, M. (1974). Microbial formation of environmental pollutants. In: *Advances in Applied Microbiology*, Ed. D. Perlman. Vol 18, Academic Press, New York; 1–73.

BOLLAC, J-M. (1974). Microbial transformaion of pesticides. In: *Advances in Applied Microbiology*,Ed. D. Perlman, Vol 18, Academic Press, New York; 75–130.

HIGGINS, I. J. and BURNS, R. G. (1975). *The Chemistry and Microbiology of Pollution*, Academic Press, London.

KADIS, S., CIEGLER, A. and AHL, S. J. (eds.) (1971). *Microbial Toxins*, in 8 volumes, Academic Press, New York.

MOREAU, C. (1979). *Moulds, Toxins and Food*, John Wiley and Sons, Chichester.

Chapter 5   Pesticides and Herbicides

CARTER, L. J. (1976). Michigan's PBB incident: chemical mix-up leads to disaster. *Science*, **192**, 240–243.

CORBETT, J. R. (1974). *Biochemical Mode of Action of Pesticides*, Academic Press, New York.

GOSSELIN, R. E., HODGE, H. C., SMITH, R. P. and GLEASON, M. N. (1976). *Clinical Toxicology of Commercial Products*, Williams and Wilkins, Baltimore.

HIGGINS, I. J. and BURNS, R. G. (1975). *The Chemistry and Microbiology of Pollution*, Academic Press, London.

MALLING, H. V. and WASSOM, J. S. (1977). Action of mutagenic agents. In: *Handbook of Teratology*, Ed. J. G. Wilson and F. C. Fraser, Plenum, New York.

MORIARTY, F. (ed.). *Organochlorine Insecticides—Persistent Organic Pollutants*, Academic Press, London.

SHARVELLE, E. G. (1961). *The Nature and Uses of Modern Fungicides*, Burgess, Minneapolis.

VARO, P. (ed.) (1976). *Pesticide Chemistry—3*, 3rd IUPAC Congress on Pesticide Chemistry, Butterworths, London.

Chapter 6 Toxic Metals

HIGGINS, I. J. and BURNS, R. G. (1975). *The Chemistry and Microbiology of Pollution*, Academic Press, London.

HMSO (1976). *Environmental Mercury and Man*, Pollution Paper No. 10, Department of the Environment, Central Unit on Environmental Pollution, London.

LUCKEY, T. D. and VENUGOPAL, B. (1977). *Metal Toxicity in Mammals*, Vol 1, Plenum Press, New York.

NORDBERG, G. F. (ed.) (1976). *Effects and Dose-Response Relationships of Toxic Metals*, Elsevier, Amsterdam.

VALKOVIC, V. (1975). *Trace Element Analysis*, Taylor and Francis, London.

WOOD, J. M. (1976). The biochemistry of toxic elements in aqueous systems. In: *Biochemical and Biophysical Perspectives in Marine Biology*, Ed. D. C. Malins and J. R. Sargent, Vol 3, Academic Press, London; 407–431.

Chapter 7 Atmospheric Toxicants

HIGGINS, I. J. and BURNS, R. G. (1975). *The Chemistry and Microbiology of Pollution*, Academic Press, London.

MCKEE, W. D. (ed.) (1974) *Environmental Problems in Medicine*, Charles C. Thomas, Springfield, Illinois.

MOORE, J. W. and MOORE, E. A. (1976). *Environmental Chemistry*, Academic Press, New York.

MUDD, J. B. and KOZLOWSKI, T. T. (eds.) (1975). *Response of Plants to Air Pollution*, Academic Press, New York.

STOKER, H. S. and SEAGER, S. L. (1976). *Environmental Chemistry: Air and Water Pollution*, Scott Foresman, Glenview.

Chapter 8   Petroleum and Radionuclides

AMERICAN PETROLEUM INSTITUTE (1975). *Proceedings of the 1975 Conference on Prevention and Control of Oil Pollution, San Francisco*, American Petroleum Institute, New York.

ARENA, V. (1971). *Ionizing Radiation and Life: An Introduction to Radiation Biology and Biological Radiotracer Methods*, Mosby, St. Louis.

BEYNON, L. R. and COWELL, E. B. (eds.) (1974). *Ecological Aspects of Toxicity Testing of Oils and Dispersants*, Applied Science Publishers, Barking.

DERTINGER, H. and JUNG, H. (1970). *Molecular Radiation Biology: Action of Ionizing Radiation on Elementary Biological Objects*, translated from German by R. P. Hueber and P. A. Gresham, Springer Verlag, Heidelberg.

INTERNATIONAL ATOMIC ENERGY AGENCY (1963). *A Basic Toxicity Classification of Radionuclides*, Technical Reports Series No. 15, International Atomic Energy Agency, Vienna.

KALSI, S. S. (1974). Oil in Neptune's kingdom: problems and responses to contain environmental degradation of the oceans by oil pollution. *Environmental Affairs*, **3**, 79–108.

Chapter 9   Assessment of Environmental Risk

BATTELLE MEMORIAL INSTITUTE (1975). *Environmental Impact Monitoring of Nuclear Power Plants—Source Book of Monitoring Methods*, two volumes prepared for Atomic Industrial Forum by Battelle, Richland and Columbus.

BONN, G. S. (1973). *Information Resources in the Environmental Sciences*, University of Illinois Press, Urbana.

CALABRESE, E. J. (1978). *Methodological Approaches to Deriving Environmental and Occupational Health Standards*, John Wiley and Sons, New York.

GOULDEN, P. D. (1978). *Environmental Pollution Analysis*, Heyden, London.

HORNE, R. A. (1978). *The Chemistry of our Environment*, Wiley Interscience, New York.

HOLISTER, G. and PORTEOUS, A. (1976). *The Environment—A Dictionary of the World Around Us*, Arrow Books, London.

KREBS, C. J. (1978). *Ecology. The Experimental Analysis of Distribution and Abundance*, Harper and Row, New York.

SOUTHWOOD, T. R. E. (1978). *Ecological Methods*, 2nd edn., Chapman and Hall, Andover.

US DEPARTMENT OF HEALTH EDUCATION AND WELFARE (1970). *Man's Health and Environment—Some Research Needs*, US

Department of Health, Education and Welfare, Washington DC.

VALKOVIC, V. (1975). *Trace Element Analysis*, Taylor and Francis, London.

VARO, P. (ed.) (1976). *Pesticide Chemistry—3*, 3rd IUPAC Congress on Pesticide Chemistry, Butterworths, London.

WINTON, H. N. M. (ed.) (1972). *Man and the Environment. A Bibliography of Selected Publications of the United Nations System, 1946–1971*, Unipub/Bowker, New York.

Appendix 1   The Relationship of Toxicity to Thermodynamic Activity

CRISP, D. J., CHRISTIE, A. C. and GHOBASHY, A. F. A. (1967). Narcotic and toxic action of organic compounds on barnacle larvae. *Comparative Biochemistry and Physiology*, **22**, 629–649.

FERGUSON, J. (1939). The use of chemical potentials as indices of toxicity. *Proceedings of the Royal Society of London*, B, **127**, 387–404.

O'BRIEN, R. D. (1967). *Insecticides Action and Metabolism*, Academic Press, New York.

Appendix 2   Environmental Indices

INHABER, H. (1976). *Environmental Indices*, Wiley Interscience, New York.

THOMAS, W. A. (ed.) (1972) *Indicators of Environmental Quality*, Plenum, New York.

US COUNCIL ON ENVIRONMENTAL QUALITY (1976). *Environmental Quality, 7th Annual Report*, US Council on Environmental Quality, Washington DC.

Appendix 3   Environmental Impact Statements

ROSEN, S. J. (1976). *Manual for Environmental Impact Evaluation*, Prentice Hall, New York.

SCIENTIFIC COMMITTEE ON PROBLEMS OF THE ENVIRONMENT (SCOPE) (1975). *Environmental Impact Assessment—SCOPE 5*, ICSU-SCOPE, Toronto, available through John Wiley and Sons, Chichester.

WARNER, M. L. and BROMLEY, D. L. (1974). *Environmental Impact Analyses: A Review of Three Methodologies*, University of Wisconsin Water Resources Center, Madison.

Guide to Information Sources

(a) Abstracts and indices.
Abstracts of Health Effects of Environmental Pollutants.
Aquatic Abstracts.
Air Pollution Abstracts and index.
Biological Abstracts and index.
Chemical Abstracts and index.
Current Contents.
Environmental Abstracts (formerly *Environmental Informaion ACCESS*) and index.
KWIC Index of Air Pollution Titles.
Pesticides Abstracts (formerly *Health Aspects of Pesticides Bulletin*).
Pollution Abstracts and index.
Science Citation Index.
Water Resources Abstracts and *HYDATA* (Water Resources Index Monthly).

(b) Periodicals.
Ambio Journal of the Human Environment, Research and Management.
American Scientist.
Archives of Environmental Health.
Atmospheric Environment.
Biologist.
CRC Critical Reviews in Environmental Control.
Ecotoxicology and Environmental Safety.
Endeavour.
Environment.
Environmental Affairs.
Environmental Health Perspectives.
Environmental News.
Environmental Science and Technology.
Essays in Toxicology, Academic Press, New York (serial publication).
Journal of the Air Pollution Control Association.
Journal of the Water Pollution Control Federation.
Nature.
New Scientist.
Oceanus.
Pesticide Biochemistry and Physiology.
Pesticide Science, Academic Press, New York (serial publication).
Residue Reviews, Springer Verlag, Heidelberg (serial publication).
Science.

Science Progress (Oxford).
Scientific American.
Water Research.

(c) Books.
GOSSELIN, R. E., HODGE, H. C., SMITH, R. P. and GLEASON, M. N.
(1976). *Clinical Toxicology of Commercial Products*, Williams
and Wilkins, Baltimore.
HMSO (1971 onwards). *Reports of the Royal Commission on
Environmental Pollution.*
HMSO (1974 onwards). *Pollution Papers.*
RUDD, R. L. (1977). *Environmental Toxicology: A Guide to
Information Sources*, Man and the Environment Information
Guide Series Vol. 7, Gale Research Co., Detroit.
SCIENTIFIC COMMITTEE ON PROBLEMS OF THE ENVIRONMENT (SCOPE)
(1971 onwards). Numerous reports. John Wiley and Sons,
Chichester.
SUNSHINE, I. (ed.) (1969). *CRC Handbook of Analytical
Toxicology*, The Chemical Rubber Company, Cleveland.
WINDHOLZ, M. (ed.) (1976). *The Merck Index*, 9th edn., Merck and
Co. Inc., Rahway.

Index